LOOKS
LIKE RAIN

9,000 Years of Irish Weather

Damian Corless, a journalist and a former editor of *Magill* and *In Dublin*, currently contributes to the *Irish Independent*. He has written comedy sketches for BBC TV's classic *Big Train* and RTÉ's award-winning *Stew*. His acclaimed books include *GUBU Nation* and *The Greatest Bleeding Hearts Racket in the World*. His most recent book, 2011's *You'll Ruin Your Dinner*, looked back fondly over Ireland's love affair with sweets down the decades.

www.damiancorless.com

LOOKS
LIKE RAIN

9,000 Years of Irish Weather

DAMIAN CORLESS

The Collins Press

FIRST PUBLISHED IN 2013 BY
The Collins Press
West Link Park
Doughcloyne
Wilton
Cork

© Damian Corless 2013
Images courtesy of the Dover Collection.

Hardback ISBN: 9781848891814
PDF eBook ISBN: 9781848898141
EPUB eBook ISBN: 9781848898158
Kindle ISBN: 9781848898165

Design and typesetting by Hugh Adams, AB3 Design
Typeset in Bembo, Foundry Sterling and Trajan
Printed in Poland by Białostockie Zakłady Graficzne SA

For Sophie, Ollie, Max and Caitlin.
Enjoy! I'll be testing you on it later.

CONTENTS

Acknowledgements   9

Introduction   11

1. Island of Perpetual Gloom   16

2. Never a Single Nail or Screw   24

3. The Dark Ages Really Were Dark   30

4. Is It A Bird? Is It A Fish?   40

5. The Answer is 42   47

6. When Murphy Says Frost, Then It Will Snow   52

7. Ghost Ships and Crystal Pillars   57

8. Waves of Mutilation   65

9. When Weather Changed History – Part 1   71

10. Rain and Slime   74

11. Weather and Witchcraft   76

12. When Weather Changed History – Part 2   80

13. Toffee for the Northern Climate   84

14. When Weather Changed History – Part 3   88

15. A Tax on Heat and Light   93

16. When Weather Changed History – Part 4   97

17. A Man For All Seasons   101

18. When Weather Changed History – Part 5   104

19. Herrings Were Found Six Miles Inland   109

20. Pills, Thrills and Weather Balloons   115

21. The Measure of a Man   123

22. The Other Potato Famine   129

23. When Weather Changed History – Part 6   134

24. A Mighty River in the Ocean   140

25. When Weather Changed History – Part 7   148

26. Death Blowing in the Wind   156

27. Force 10 at Fastnet   160

28. The World's First Weather Forecast (Gets It Right!)   168

29. Weather and Wireless   173

30. Below Us The Waves   180

31. Clocks and Cows   189

32. The Farmer and the Fisher Should Be Friends   196

33. When Weather Changed History – Part 8   202

34. No Respecter of Fame   207

35. Night Appears To Fall By Midday   212

36. The Minister for Snow   217

Selected Bibliography and Sources   222

ACKNOWLEDGEMENTS

Thanks to The Collins Press, Faith O'Grady, Michael Gallagher, Professor Mike Baillie and Hugh Adams.

INTRODUCTION

'Everyone talks about the weather but nobody does anything about it.'

That witticism is commonly attributed to Mark Twain, although, like many bon mots attributed to the writer, he may never have said it. Of course, while the quote is in one sense a statement of the obvious, it's not entirely true. The history of humankind is very much a history of trying to do something about the weather, from wrapping up in furs to ward off the cold to seeding clouds with chemicals to make it rain.

In Ireland, unlike lands with much more predictable patterns, the first obstacle to doing something about the weather is that we never really know what the weather is going to do. The Dublin writer Oliver St John Gogarty told of how he'd remarked to a man that the island was experiencing 'the most extraordinary weather for this time of year'.

The man corrected him: 'Ah, sure it isn't this time of year at all.'

On a similar note, making small talk in the film *Way Out West* Oliver Hardy ventured: 'A lot of weather we've been having lately.'

Ireland has a lot of weather. Grappling with four seasons in one day is a mundane condition of living here.

Except, of course, once we give it any thought we appreciate that the weather is anything but mundane.

On the contrary, the weather is one of the great unknowns – and arguably unknowables – facing all life on Earth.

Unknowable it may ultimately be, but that has never stopped people from trying to get to know it or – failing that – letting on to know it.

Attempting to put some sort of order on disorder is a primal human impulse. It is essential to our happiness and indeed to our sanity to feel we have some degree of control over the world about

us. Big and scary and contrary, the weather has never hesitated to punish that sin of pride.

From the time of the ancient Babylonians, and probably long before, the art of weather divining has attracted more than its share of chancers and charlatans, along with earnest proto-scientists like Aristotle. For all its many faults, the Greek philosopher's *Meteorologica* (*Meteorology*) remained the standard textbook on climate for some 2,000 years after its publication around 340 BC.

But even as the new scientists of the Enlightenment were picking apart Aristotle's theories, the biggest thing in weather throughout the eighteenth century and into the nineteenth was the hugely popular almanac. Scores of these publications in the Old World and the New purported to carry specific forecasts into the distant future. The great Benjamin Franklin published one of America's best-read almanacs, while the not-so-great Corkman Patrick Murphy scored a runaway bestseller in the British Isles on the back of one blessedly fluke forecast. (See pp. 51–52.)

The appliance of science to weather forecasting has not always been appreciated by everyone. Indeed, it was – and remains – a frequent subject of scorn. When the visionary weatherman Robert FitzRoy set up the world's first truly scientific network of weather-reporting stations around Britain and Ireland, the almanac publishers and shipowners led the clamour for his head on a plate and his system to be demolished. The former correctly feared that scientific forecasts would ruin their bogus business, while the latter were angered when sailors refused to put to sea in heed of FitzRoy's storm warnings.

Today, for all the mod cons at their disposal, our professional forecasters regularly find themselves falling back on an age-old practice, described by the late doyen of Irish meteorology Brendan McWilliams as 'the honourable ploy of hedging'. This is the practice whereby forecasters 'hedge' their bets by predicting every type of weather for the coming period in the hope of bluffing through.

While we have gotten the measure of our weather to a much greater extent over the past 300 years, it can still exert an almost supernatural presence in our lives. Like the Greek gods of old it is prone to the wildest mood swings, on a seeming whim lashing out in a huff or coming on all lovey-dovey.

The Victorian critic John Ruskin fabricated the term 'pathetic fallacy' for instances where we project feelings, motivations or intentions onto non-living things. The word 'pathetic' relates to 'pathos' or 'empathy' and isn't meant to suggest that the delusion is wretched, pitiable or lame. In an exclusive interview for this book, Professor Mike Baillie of Queen's University, Belfast, suggests that a close encounter with a comet long ago gave birth to the Cúchulainn myth, as the ancient Irish projected a superhero personality onto this terrifying apparition in the sky.

There can be no doubt that the weather is part of what we are, although there is great debate as to the size of that part and where it is to be found. There is sound scientific evidence, for instance, that the profusion of red hair and freckles in the Irish is a product of our sun-starved climate.

As for its effects on our national character (if we have one), on our alleged tendency to melancholy (if that's true) and on our drinking, the jury is locked and is unlikely ever to agree on a verdict. The poet Wallace Stevens insisted: 'The state of the weather soon becomes a state of mind.' This view had already been considered and rejected by the literary maestro Samuel Johnson who rubbished the notion that the mood or mindset of a rational being could be in any way moulded by the weather. Johnson wrote disdainfully: 'Surely nothing is more reproachful to a being endowed with reason than to resign its powers to the influence of the air, and live in dependence on the weather and the wind.'

For what it's worth, after half a century living in, and with, the Irish weather, I think Johnson's attempt to downplay so absolutely its influence on our world view and well-being is daft.

This book explores the profound impact of the weather on the Irish people since the first settlers arrived some 9,000 years ago. The main lesson I've learned in that exploration is that for roughly 8,900 of those years the actions of the weather were a desperately real matter of life and death. Going back just a very few generations, our agrarian ancestors lived precariously in the full knowledge that they were just one bitter winter or cruel summer away from starvation, pestilence and flight or death.

The other lesson learned is that while we have insulated ourselves from our weather to an impressive extent, we have never come near to taming it.

And I don't believe we ever will.

Damian Corless, 2013

ISLAND OF PERPETUAL GLOOM

A Warm Welcome
for the First Settlers

To the Arabs of North Africa a thousand years ago, Ireland was a cold, forbidding place jutting out of the tempestuous and fearsome ocean they called the Sea of Perpetual Gloom. According to the twelfth-century geographer Al Idrisi, the few ships that left the placid Mediterranean to venture northwards did so only during the 'favourable (summer) season, as soon as the weather is calm and the Sea of Perpetual Gloom is tranquil'.

A thousand years earlier still, the Romans had kicked around the notion of invading Ireland but then thought the better of it.

Agricola, a Roman governor of Britain in the first century AD, calculated that a force of 6,000 crack troops could subjugate the entire island, but either he or his superiors developed cold feet.

Cold feet may have been a key factor in the decision to leave the Irish to their own devices. The name the Romans gave to Ireland was Hibernia, meaning 'Land of Winter'. The Roman name seems to be a mistranslation or corruption of the Greek word 'Iverni' which referred to the people of the southern tip of Ireland. The Greek geographer Strabo claimed that the miserable climate was responsible for producing generation upon generation of cruel, cannibalistic savages who generally cursed their misfortune at being born in such a godforsaken place.

For the Greeks, the Romans and the Arabs settled around the sun-kissed Mediterranean, Ireland must indeed have seemed like the last place on Earth that anyone would choose to live. But choose to live there people did, despite the weather. Indeed, to the first arrivals, the Irish weather probably seemed very welcoming indeed.

As the ice sheets rolled back at the end of the last Ice Age around 9,000 years ago, the first people set foot in the newly uncovered Ireland. Who they thought they were, what they called themselves, and where in Britain or Europe they'd originally set out from is a matter of some dispute.

The first settlers arrived around the beginning of a particularly warm period known as the Holocene Climatic Optimum. This pleasant interlude kept Ireland up to $3°C$ warmer than it is today for the first 4,000 years of human habitation, spanning from around 7,000 BC to 3,000 BC.

The first known settlements are concentrated in the northeast of the island, which leads some experts to believe that hunter-gatherers simply strolled across a narrow land bridge from Scotland which had yet to be covered by rising melt waters. The course of this land bridge would roughly coincide with today's popular ferry route between Larne on the Antrim coast and Stranraer in Scotland.

A to Z
of Irish Weather

Air Thermals A thermal is a column of rising air caused by the uneven heating of the Earth's surface. These often rise above cities and towns, signifying 'heat islands' which trap more warmth in stone and concrete than is retained in less built-up surrounding areas. In 2012 a study by NUI Maynooth, which monitored fifteen weather stations for two years in the greater Dublin area, found that the night-time temperature in the centre of the capital was on average between 4 °C and 8 °C warmer than in the surrounding suburbs. By day it was 1.5 °C warmer.

Anemometer The anemometer is a device for measuring wind speed and is commonly found in weather stations. The first surviving description comes from Italy around 1450, but the best-known type, the cup anemometer, was invented by Dr John Romney Robinson of Armagh Observatory in 1846. The Robinson Crater on the Moon is named after him, while his daughter married the Irish physicst George Stokes who conducted groundbreaking work on light, rainbows, clouds and waves. *See* Stokes' Law.

Anticyclone The opposite of the familiar, unsettled low-pressure cyclone, which rotates anticlockwise above Ireland (as it does in all of the northern hemisphere), the high-pressure anticyclone rotates clockwise and is associated with dry, clear weather.

Augurs These forecasters of Ancient Rome had a wide-ranging remit which included predicting the weather. They watched for auspices of things to come, and paid particular attention to the behaviour of birds. The augurs of Celtic high society ranked below druids but above bards.

April April was the second month of the ten-month Roman calendar before January and February are said to have been added around 700 BC by King Numa Pompilius, who may or may not have really existed. Despite tweaks to the Western calendar, it has remained the fourth month ever since.

August The eighth month of both the Julian and Gregorian calendars, August was named after Augustus, the ruler who transformed Rome from a Republic into an Empire with himself as the first Emperor.

Depending on whether the climate was blowing hot or cold, the Scottish–Irish land bridge may have sunk below the waves and resurfaced a number of times in the 3,000 years before it finally disappeared around 9,000 years ago, when global warming finally turned Ireland into an island.

Some geologists believe that there were several trackways joining Ireland to Britain stretching from Antrim in the north to Waterford or Cork in the south. These would account for the fact that red deer, wild boar and other large immigrant animals had penetrated independently into the midlands, which were cut off from the first settlements by distance and thick forest.

This theory of southern land bridges is also favoured by those who believe that Ireland was initially settled by groups of Iberian beachcombers who followed the coastline newly exposed by the retreating ice from northern Spain to Ireland, grazing on relatively easy pickings along a thousand-mile shoreline smorgasbord of shellfish, seabird eggs and other convenience foods.

Ireland became an island long before neighbouring Britain, which retained its umbilical cord to Europe for centuries more. Ireland's much earlier isolation meant that many varieties of plants and animals which reached Britain never crossed the Irish Sea. It is estimated that some 30 per cent of the plant and animal species that are native to Britain, snakes being an obvious example, never reached Ireland.

For every expert who leans towards the land-bridge theory as the gateway for Ireland's first post-Ice Age settlers, there's another who argues that they most probably arrived by boat. The land exposed by the retreating ice was quickly covered by tundra composed of low-lying shrubs, sedges, mosses, grasses and lichens. But this in turn quickly gave way to thick forests which blanketed Ireland, Britain and most of Europe. For anyone wanting to get from Point A to Point B, it was often much quicker and safer

What To Wear 7,000 BC

Keeping Warm and Dry

There's an old saying that there's no such thing as bad weather, only bad clothing. This, of course, is only half right. The first people to settle in Ireland some 9,000 years ago will have encountered plenty of bad weather, but we can be sure that they arrived in clothing that was well tailored for the worst the elements could fling at them.

Because of Ireland's damp climate, and the perishable nature of clothing materials, we have only fragmentary evidence of how Ireland's first populations dressed to keep warm and dry. For an indication of the type of clothes they might have worn, we can look to Ötzi the Iceman, a mummy discovered in 1991 in the Tyrolean Alps separating Austria and Italy.

Although the Iceman lived around 3,300 BC, much later than Ireland's first settlers, his well-preserved clothes will have differed little from those of the people who repopulated the continent as the ice sheets retreated. These people were well capable of making garments that were sophisticated, snug and superbly fit for purpose. And they had to be, in a time when the penalties for bad clothing could be far more lethal than ending up on a 'Worst Dressed' list.

Ötzi's clothes were comfortable, warm and hard wearing. Apart from a cloak made of woven grass, his garments were entirely made from the skins of various animals. In addition to a leather coat, a belt, a pair of leggings, a loincloth and shoes, he also wore a bearskin cap with a leather chinstrap. The shoes were broad, waterproof and designed for walking across snow. They were made using bearskin for the soles, deer hide for the upper panels, and a stocking-like netting made of tree bark. The feet were well padded with soft grass and moss that keep them as snug as a pair of socks. Ötzi's coat, belt, leggings and loincloth were fashioned from strips of leather sewn together with sinew. His utility belt had a pouch sewn to it containing a cache of handyman's tools, including a scraper, drill, flint flake and a bone awl. It also contained dried fungus in what may have been a medicinal first-aid kit.

By the time Ötzi met his untimely end in the Alps, farming had been established in Ireland for around 700 years, providing the population with a yearly yield of wool from sheep and goats. The wool was often dyed before spinning and then woven into warm cloth. There is some dispute about when Irish flax was first processed into linen, with many experts dating it to the first century AD. Tanning and leather embossing were known from early times, and leather bags served as canteens for water before pottery became widespread. Simple slipper-like rawhide shoes made from a single piece of skin have been found, along with more sophisticated designs.

Over the millennia, from the arrival of the first people to the coming of the Anglo-Norman scribes who studied the natives with disdain, the fashions of the Irish evolved. One key aspect of the native dress which remained constant was the use of layers to keep out the wind, the rain and the cold. Multiple layers of clothing trap pockets of air which quickly absorb warmth from the body. These warm layers in turn keep the body insulated against the elements.

The Anglo-Normans, who arrived in the twelfth century, noted that the Irish wore a wide assortment of furs, including otters, seals, foxes, wolves, badgers and other wild animals. These highly valued furs were used as the first line of defence against the wind and the rain.

Generations of Irish schoolboys and schoolgirls were taught that the crowning glory of Irish clothing design, in terms of form and function, was the Irish Mantle. Some sources even refer to it as The Great Irish Mantle, to convey its tent-like qualities.

One influential eyewitness who thought the Irish Mantle was something short of great was the leading Elizabethan Edmund Spenser, one of the most illustrious figures in English poetry. Spenser spent much of his later life in Cork during the period of the Munster Plantation. It was there that he made friends with his neighbour and fellow planter Sir Walter Raleigh, and there where he is said to have composed sections of his groundbreaking masterpiece *The Faerie Queen*.

It would not do Spenser's memory a disservice to say that he feared and loathed the native Irish. In his 1596 pamphlet, *A View of the Present State of Ireland*, he argued that the land would never be fully 'pacified' until Irish ways and Irish laws were purged from the face of the Earth. In fact, his solution to England's 'Irish Problem' was nothing short of genocide.

At best, Spenser could muster a grudging respect for the all-purpose, all-weather Irish Mantle. He wrote: 'It was their house, their tent, their couch, their target [shield]. In summer they wear it loose, in winter wrap it close.' Passed in 1366, but more observed in their breach than their observance, the Statutes of Kilkenny banned the English settlers from going native. This ban, which specifically forbade the wearing of the Irish Mantle, was supposed to save them from degenerating to the level of the savage Irish.

In Spenser's *A View of the Present State of Ireland*, he presents the mantle as a symbol of decline and descent into bestiality. He asserts that the mantle started out as the noble cape of ancient Egypt, Israel, Greece and Rome until with 'the decaie of the Romaine Empire' it fell into the hands of barbaric societies like the Gauls of northern Europe and the Sythians of the Near East. The poet went to great length to denigrate the Irish Mantle as the camouflage of a sneaky fighter and the pop-up bed of a prostitute. Other agents of Elizabeth's rule in Ireland weren't put off by this black propaganda. In 1599 one English army quartermaster submitted a request for a shipment of Irish Mantles for his wet, cold and miserable troops, noting that the native garment was far superior to anything the English had in their wardrobes. Shortly after the English victory over a combined Spanish/Irish force at the 1601 Battle of Kinsale, Queen Elizabeth I posed for her famous Rainbow Portrait in which she reasserted her dominion over Ireland by draping herself in a multicoloured Irish mantle.

to go by boat than to beat a trail through the dark, dense, scary woodlands populated by wolves, bears and evil spirits.

As mentioned, the climate encountered by Ireland's first settlers was reasonably mild. The summers were rarely scorchers, but equally the winters were mostly tolerable. The first settlers no doubt regarded the wishy–washy Irish weather as a mixed blessing, but they would have been wise enough to appreciate that its absence of extremes was a boon. Its general lack of a vicious streak was a deeply attractive quality to people who were acutely aware at all times that they existed precariously just one cruel winter or one failed harvest away from the next life.

Perhaps remarkably, Ireland's weather has remained that same moderate mixed blessing down to the present day, with very few deadly twists and turns over the course of the past 9,000 years. There have been some extreme episodes of prolonged famine and pestilence down the millennia, but the rarity of such fierce disruptions serves to underline just how stable and dependable Ireland's climate has been for its people.

For comparison, let's take North Africa. As the first hunter-gatherers arrived in Ireland around 9,000 years ago during the Holocene Climatic Optimum, the landscape of North Africa was being transformed beyond recognition by the Holocene Wet Phase. As humans began pushing into Ireland, the Sahara was being swamped with wave upon wave of new settlers, as monsoon rains transformed the North African plain from a featureless desert into a lush, green paradise swarming with wildlife.

Some 3,000 years later, when the first farmers arrived in Ireland with their packages of seeds and domestic animals, the Irish landscape had altered but the weather which greeted them from day to day and year to year wasn't much different from that encountered by the first arrivals, and not much different either from what we put up with today. By the time farming was becoming widespread across Ireland, however, North Africa was experiencing another massive climatic upheaval. The flow of monsoon rains diverted southwards, and humans were driven off the Sahara as it rapidly reverted to the harsh, lifeless desert it had been before.

NEVER A SINGLE
NAIL OR SCREW

The Thatched Roof

THE HISTORY OF IRISH HOUSING is the history of keeping out the rain, and the thatched roof has been part of that history from the time people first set foot on this island, or not long after.

Weaving or layering plant material into shelters has been around from the dawn of mankind and remains one of the most popular forms of roofing in the world today. Historians believe that the first hunter-gatherers to arrive in Ireland around 9,000 years ago would have been skilled in the techniques of making fairly watertight roofs from a thatch of broom, rushes and reed, plugged with heather, mosses and mud. It may be, however, that the very first explorers carried their own camping gear comprised of animal skins stretched over poles, and that these skin shelters served very well for some time.

Around 4,000 BC the Neolithic agricultural revolution arrived in Ireland, bringing a package of new cereals, including wheat and barley, together with new fashions in housing. While the king of the new cereals, wheat, settled uncomfortably into the damp Irish climate, barley and rye fared better. Grain and chaff made up around half the volume of the new cereal crops. Once that half had been separated out, the 50 per cent left was straw, consisting of the dried stalks. The many uses of straw included brick-making, rope-making, animal feed, bedding, basket weaving and thatching.

After thousands of years as the most indispensible roofing material in the land, thatch began to fall out of favour in the early years of the twentieth century. From the 1920s the new Free State government began providing grants for home improvements, and in 1936 the Minister responsible, Sean T. O'Ceallaigh, told the Dáil with evident satisfaction: 'Concrete floors are substituted for earthen floors, and timber work renewed as required. In some areas thatch roofs, which necessarily require attention frequently, are being removed and replaced with either native slates or concrete tiles of Saorstát (Free State) manufacture with the assistance of the reconstruction grant. The low, unsightly and unhealthy corrugated iron structures which hitherto were to be seen in some of our rural areas are also disappearing and are being replaced by well-ventilated rooms of adequate height with slated or tiled roofs.'

The main arguments of the Free State's modernisers against thatched roofs were that they leaked, that they had to be replaced every few years, that it was becoming ever harder to secure the services of a skilled thatcher, and that when you could get a thatcher, he'd charge you an arm and a leg. Traditionally the skills and trade secrets of the craft had been handed down from father to son, but the flight from the land, and from the island, was thinning their ranks dramatically. And thatching wasn't something that could be picked up easily from a DIY manual. As one writer pithily observed: 'Trying to explain to a lay person how to thatch a house is like describing to someone who has never worn a shoe how to tie a shoelace. Very nearly impossible.'

For the record, the straw thatch of a new roof had to be first stitched into the frame with a needle two feet long. It was then fastened down hard with hazel and willow twists. A proper thatcher would never use a single nail or screw to secure it to the joists. Rethatching an existing roof, meanwhile, required another skill set.

While the Fianna Fáil housing minister O'Ceallaigh was preaching the advantages of slated and tiled roofs, *The Irish Press*, the newspaper set up as his party's mouthpiece just five years earlier, was running a counter-campaign to save Ireland's thatched roofs.

A TO Z
OF IRISH WEATHER

Bank Holidays In 2012 Gerald Fleming of Met Éireann revealed on the TV documentary *Weather Permitting*: 'Bank holiday weekends are a forecaster's nightmare. Expectations are sky high – levels are in the heavens. No one likes working on a bank holiday because you're on a hiding to nothing if you get it wrong.'

Baths The hot bath was a favourite means of easing away the worst effects of the Irish weather in Celtic society. Luxuriating in a hot bath is a recurring boastful feature of tales detailing the lifestyles of the rich and famous in ancient Ireland. Every monastery kept a bath in its guest house for distinguished visitors.

Bealtaine The ancient Celtic May Day festival to celebrate the waxing power of the sun, marked with bonfires and feasting.

In a 1932 feature headlined 'Will The Thatchers Die?', the *Press* put the decline in thatching down in large part to snobbery and what it felt was the decidedly un-Irish trait of keeping up with the Joneses.

The paper lamented: 'One of the causes for the decline of thatching was a silly notion of inferiority. People thought that it was backward to live under thatch; they thought that they went up the social scale when they put slate or tiles over their heads.' Arguing that 'wherever there are thatched houses there is beauty', the *Press* writer added: 'Not that the rugged Irish slate looks badly. It is more

Beaufort Scale Properly titled the Beaufort Wind Force Scale, this system devised by Meath man Sir Francis Beaufort measures wind speed by relating it to observed conditions on sea or land. (See pp. 119–124.)

Blizzard A very cold, strong wind (Force 7 or above on the Beaufort Scale) laden with snow.

Boyle, Robert Born in County Waterford in 1627, Boyle was a natural philosopher, physicist and chemist and is best known for devising Boyle's Law which defines the relationship between the pressure of air and its volume. Amongst other things, Boyle's Law is the principle governing why weather balloons are only partially inflated at launch. As the balloon gains altitude, the air pressure decreases and the gas expands. Being partially inflated at lift-off enables the expanding balloon to reach its intended altitude without mishap and remain aloft long enough to gather the desired data.

Bus Shelter The bus shelter has been a symbol of urban decay and moral rot since it first appeared in Irish towns, providing the punch line to the joke: 'What do Northsiders use for protection during sex?' In 1946 a scheme to provide many of Dublin's main routes with bus shelters was abandoned due to wanton vandalism after just twenty-two had been erected. The glass in many was smashed, while in some cases thieves swiped the large panes intact for recycling as windows. A letter writer to *The Irish Press* that year suggested that the shelters were too good for the uncouth Irish public who would be better served with 'sturdy corrugated iron structures instead of glorified glasshouses'.

agreeable to the eye than the sleek Welsh slate. As for tiles, now being introduced in many places, they are a miserable commercial imitation of a thing which is beautiful enough in its own county, but alien and inharmonious when brought to us. Cotswold tiles are one of the graces of a peculiar West of England architecture. The imitation tile in Ireland is a disgrace.'

Two years later, in 1934, *The Irish Press* was still waging its losing battle to persuade the homeowners of Ireland that thatch was best. It reported a widespread disillusionment with the new style of house popping up everywhere, insisting: 'When the medical officers visited the people in their up-to-date residences, complete with all modern conveniences, they were met with a barrage of complaints. Women alleged that they and their families were "perished with the cold", that all the windows made the place draughty, and that the "old thatched cottage with all its faults was a palace compared to the damp new house".' The paper said that a skilled thatcher 'will take a week to roof a small dwelling at a cost of around £5. A large house may cost £30.' The *Press* conceded that the modern thatch didn't last quite as long as the old thatch, because mechanised threshing techniques 'break the barrel of the straw'. Even using machine-damaged straw, a new thatched roof would keep out the rain for four or five years.

By 1970 even *The Irish Press* had given up on its hopes that the thatched roof could be saved as a quaint feature of the Irish landscape. That year Galway County Council appealed to Local Government Minister Bobby Molloy for a special grant to replace thatched roofs with slate or tiles. It reported: 'Of the 2,961 "unfit" houses in the county the large majority are thatched cottages which have fallen into disrepair and for which thatchers cannot be got.'

As the second big oil crisis of the 1970s hit hard in 1979, *The Irish Press* reprised an old tune with an article suggesting that 'a good roof of thatch is one way to beat the predicted winter fuel shortage'. In reality, though, the piece was a requiem for a dying craft. It focused on a veteran Kildare thatcher named Paddy who knew he was amongst the last of his breed. 'There's no young lads going into the business,' he reflected, adding that good straw was virtually impossible to come by. He remarked: 'They put too much fertiliser on it now and it comes up too fast, and that makes it too soft. Also the modern cutting machinery makes a mess out of it.'

Decades earlier, when Paddy began learning the craft from his father, 'there were more thatchers in Kildare than there are today in the whole country, and competition was fierce'. In the space of Paddy's lifetime, the thatched roof has passed from an age-old basic necessity to a picture-postcard prop.

THE DARK AGES REALLY WERE DARK

Irish Oak, Comets and Catastrophes

E VERY SCHOOLCHILD KNOWS (although not every scientist agrees) that the dinosaurs were wiped out 65 million years ago by an asteroid strike off the Mexican coast. The impact of the missile from space is thought to have cloaked the planet with a thick veil of soot and debris, blocking the sun's light and heat for a long, long time.

The term 'nuclear winter' was coined in 1983 by a team investigating the disastrous after-effects that a nuclear exchange would have on the planet's weather for years after the launch buttons were pushed. Its meaning has since been expanded to include climactic catastrophes such as the one that ended the reign of the dinosaurs, and the cold Dark Age caused by Sumatra's super-volcano some 74,000 years ago that some argue brought early *homo sapiens* to the very precipice of extinction before we'd really got going.

There have been other nuclear winters much closer to our own time, although none on anything like the same scale. The causes of these Dark Ages have occupied the mind of Mike Baillie of Queen's University, Belfast, for much of his career. As the University's Emeritus Professor of Palaeoecology, and one of the world's leading authorities on tree-ring dating, Professor Baillie has spent decades using fossils to reconstruct the climate and ecosystems of the ancient past. His researches have led him to believe that Ireland, Europe and perhaps the entire globe was plunged into nuclear winters several times over the past few thousand years, sometimes by the eruption of supervolcanoes and sometimes by comets passing a little too close for comfort.

Using ancient tree-ring evidence from Ireland and beyond, cross-referenced with ice-core drillings from Greenland and other data, Mike Ballie and like-minded experts make a persuasive argument that the human race has had to survive a number of severe nuclear winters in the relatively short time since the first settlers arrived in Ireland around 9,000 years ago. Allowing for a slim margin of error on either side, these abrupt downturns in the weather can be pinpointed with remarkable accuracy to 4375 BC, 3195 BC, 2345 BC, 1628 BC, 1159 BC, AD 207 and AD 540. Few scientists dispute that these downturns happened at the given dates, but many insist that these snaps of nuclear winter must all be put down to supervolcanoes. Ballie and others argue that in some cases there's little or no evidence for volcanic disruption, and comets can be placed in the frame as possible culprits.

The most useful and accurate tool we have for charting the climate of the past is tree-ring dating, or dendrochronology. The science was developed in the United States by the astronomer A. E. Douglas, who found that he could link the growth of tree rings to the amount of sunlight and warmth in the air. His equation was devastatingly simple: fine weather meant thick rings and harsh weather meant thin rings.

After a sluggish start, dendrochronology took off in the 1960s and today provides a fully anchored (that is, reliably accurate) chronology of the changing state of the planet's climate over the past 11,000 years. Preserved in pristine condition, the data gathered from oak and pine trees recovered from Irish bogs forms a major plank of this rich body of evidence.

Mike Baillie explained to this author that, in the late 1960s, Queen's University invested in laboratory equipment to carry out radiocarbon dating. This was a relatively new discipline which allowed carbon-bearing materials (plants, fossil fuels like coal, etc.) to be dated with some accuracy back 60,000 years or so. At first, the Queen's scientists concentrated on dating pollen samples preserved in the bogs, until it dawned on them that there were much bigger packages of climate data just waiting to be unwrapped.

Ballie recalled: 'The guys out collecting pollen samples noticed that there were lots of bog oaks being dragged out by farmers in drainage operations, so they saw this mass of material sitting in the countryside. There was a major controversy going on at the time. In order to check radiocarbon, the Americans had used very long-lived tree ring patterns, mostly from bristlecone pine which can live for 5,000 years. When they did that, the calibration curve they produced had a lot of wiggles in it.

'The controversy was: are these wiggles just statistical noise, or are they real geophysical changes in the amount of cosmic radiation hitting the earth? And the answer wasn't known, so a bunch of characters already working in palaeoecology, principally Professor Alan Smith and Dr John Pilcher, came up with the idea of building an Irish tree-ring chronology and doing an Irish calibration. And I happened to be passing the door, looking for a foothold in archaeology.

'It wasn't built with any thoughts of climate, or weather, or extreme events or anything else. However, it quickly became apparent that the only species to use if you're going to build a long chronology is oak, because the oak is there as living trees, and in historic buildings, and on historical sites, and there are all these preserved bog oaks, which had grown over the surface of bogs and then been blown over and buried. We had a huge repertoire of oaks and time, so we got stuck in. It took fourteen years to do, by which time we had a chronology back to 5,289 BC.

'The thing about dendrochronology is that you're overlaying so many ring patterns you can iron out any problem ring, and because oaks are so long-lived you can get long sections of chronology. The other joy of the method is that you can check your results against the results of independent workers building a similar chronology in Germany. So we could not only produce a chronology but we could ensure it was precise.'

By 1982 the Queen's University team and their German counterparts had gathered enough tree-ring data to publish an accurate picture of the climate over northern Europe stretching back 7,000 years. 'At that point we were out of a job,' jokes Baillie.

But before too long the Irish and German teams were called back to the front line. In 1984 a pair of American scientists published a paper based on Californian tree rings which suggested that a major downturn in the weather of the Northern Hemisphere might have been caused by the volcanic explosion of the Greek island of Thiera. Better known as Santorini, Thiera is today a popular tourist destination. In the ancient world it was one of the commercial hubs of the Minoan civilisation which appears to have had its capital on Crete, and which crumbled soon after the explosion.

There was one problem. The American research pinpointed the catastrophe at 1627 BC, but most archaeologists placed the destruction of Santorini more than a century later. Mike Baillie cross-referenced the Irish and Californian tree rings and, in his own words: 'Lo and behold, you had this extreme downturn and the very first sign of it in our Irish trees appeared to be 1628 BC'. (In BC counting, 1628 is the year before 1627, suggesting the extreme weather event hit Ireland before America.)

After investigating the weather shock of 1628 BC, Baillie began looking for clusters of very narrow growth rings going back thousands of years. His findings showed that Ireland was hit by periods of dreadful weather around – you've guessed it – 4375 BC, 3195 BC, 2345 BC, 1159 BC, AD 207 and AD 540.

The most recent of these, around AD 540, suggests that the Dark Ages really did feature a prolonged period of darkness. Baillie now believes that the AD 540 event was most probably the result of two supervolcanoes erupting a few years apart, and that the combined eruptions spewed soot and debris into the upper atmosphere which caused famine, unrest and upheaval from Peru to China, where falling snow in August damaged the harvest. The Irish Annals of this period record 'a failure of bread' in AD 536, 'a great mortality' in AD 541 and, in AD 543, 'an extraordinary universal plague through the world which swept away the noblest third part of the human race'.

In his books, Baillie has suggested that the extreme weather which ravaged the land at this time may have persuaded the Irish people to forsake their old Celtic gods in favour of the new Christian religion. He has pointed out that the people seem to have seamlessly transferred many of the superpowers of the old gods to the superheroes of the new religion, most notably Saints Patrick, Brigid and Columcille. He has also produced a graph that shows how the traditional foundation date given for Ireland's Christian monasteries spikes spectacularly between AD 540 and AD 550, the worst years of the nuclear winter. Tellingly, the ancient worship site of Tara was abandoned just as the monastery-building programme was at its most fevered.

Baillie has identified 1159 BC as the hub of another weather catastrophe, which may have spelled the destruction of an ancient way of life in Ireland and brought a thriving modern one crashing down in Greece. His examination of bog oaks from this period show that the trees in question endured eighteen summers during which they put on no growth.

By way of a health warning, he points out that trees that grew on (and fell into) poor bogland suffered a longer period of stunted growth than ones taken from ancient buildings, which more likely grew on drier, more fertile land. This indicates that the bleak weather may have hit some Irish districts and communities harder than others.

That said, Baillie asserts: 'There's little doubt that 1159 BC is clearly a big event. It shows up widely in tree rings and it's also not far from the beginning of the Greek Dark Ages. At this time there's a collapse in civilisation in the Mediterranean area.

'How bad was this? Did this effect agriculture? We do have hints of some hill forts being built in Ireland at this time. The problem is that archaeologists can't date monuments as well as we can date tree rings, so we're asking questions the archaeologists find difficult to answer. There are hints that there may have been a flush of hill fort building around the twelfth century AD, but archaeologists who don't like that idea argue that the dates could be stretched to around AD 1000.'

Most hill forts consist of lines of defensive earthworks, stockades and ditches, built around the natural contours of a hill. Logic suggests that their construction was a response to some new challenge or threat.

Baillie speculates that the new hill forts might indicate 'a centralisation of control in some areas'. He expands: 'It may well have been, with land under stress, that land where you could produce food would become more valuable and hence there was more need to defend it. This is speculation, but once you've got a big event and some archaeological phenomena, you're at liberty to speculate about how might it have played out.

'The other speculation is, what caused this downturn? It took the ice-core workers a long time to get a clear picture of what happened in the sixth century AD. For years they didn't have good evidence of those two volcanoes I mentioned. They only found that they had these big volcanic signals because people like me were prodding them. As a result of this prodding they reanalysed three ice cores and found they had these big signals that they'd previously underestimated, which is a bit alarming from a scientific viewpoint!'

While research across a spectrum of disciplines has identified two massive volcanic eruptions as the probable cause of the AD 540 global event that wiped out one in three Europeans, science has been unable to match the nuclear winter of 1159 BC with any volcanic activity. This has prompted Baillie and others to look for an extraterrestrial cause.

He says: 'There's no evidence at all of big volcanoes in that immediate vicinity. That allowed me to speculate that, if there's no big volcano, what's the next most likely explanation of a big global event where the historical record says the sun was dimmed, implying that the atmosphere was loaded with something forming a dust veil? There's two ways to create a dust veil. One is to eject stuff up into the atmosphere from a volcano, the other is to dump stuff in from space, most likely from comet debris. That goes down like a lead balloon with archaeologists and historians who do not like the concept of stuff falling out of the sky. It's called special pleading and they won't accept it unless you produce cast iron and one hundred per cent gold-plated evidence. So basically I can say what I like and they ignore what I say, which is fine because I can't absolutely prove that some events had an extraterrestrial component. But that doesn't mean that in the fullness of time some research student won't look down a microscope and ask "What are all these metallic spherules I'm finding?" – who knows?'

In support of his extraterrestrial theory, Baillie and his collaborator Patrick McCafferty have argued that close encounters with passing comets may have both shocked and awed the ancient Irish into worshiping them as gods.

The pair quote a paper by the scholar Dorothea Kenny who argued that the ancient Irish god-king Balor was a comet. She wrote: 'I suggest that a comet, Balor of the Evil Eye, broke apart – perhaps into two pieces, Lug and Balor, and that its last […] appearance, with loud noises, atmospheric dust and falls of rock

A TO Z
OF IRISH WEATHER

Celsius/Centigrade A scale and unit of measurement for temperature, named for the Swedish astronomer Anders Celsius (1701–1744) who proposed it. Centigrade is derived from the Latin term for 'one hundred steps', reflecting its 100-degree scale from freezing to boiling.

Chimes Wind chimes have been a popular garden ornament for millennia, much beloved of the ancient Chinese, Indians and Romans. In a dispute between neighbours which landed before an Irish judge in early 2013, one side claimed that noisy chimes had been deliberately used as an instrument of 'torture'.

Church Bells In medieval times the ringing of church bells was held to ward off the worst effects of storms, and especially hail which could devastate crops in the fields. The seventeenth-century Pope Urban VIII (the one who summoned Galileo to Rome to recant his work which removed the Earth from the centre of the universe) gave his imprimatur to a prayer to be used by bishops in the consecration of bells. It went: 'Grant, O Lord, that the sound of this bell may drive away harmful storms, hail and strong winds, and that the evil spirits that dwell in the air may by Thy Almighty power be struck to the ground.'

and fiery stuff, has been given romantic shape in the epic tale *Táin Bó Cuailnge*, centring on the extraordinary character Cúchulainn.'

Baillie, McCafferty and Kenny argue persuasively that several descriptions of the shape-shifting superhero Cúchulainn in the *Táin Bó Cuailnge* make a very close fit for a comet. One passage says: 'The first warp-spasm seized Cúchulainn, and made him into a monstrous thing, hideous and shapeless, unheard of … Malignant mists and spurts of fire flickered red in the vaporous clouds that rose boiling above his head, so fierce was his fury. The hair of his head twisted like the tangle of a red thorn bush stuck in a gap … Then, tall and thick, steady and strong, high as the mast of a noble ship, rose up from the dead centre of his skull a straight spout of black blood darkly and magically smoking.'

Another description from the *Táin* also evokes a flaming, violent apparition closer to a fireball in the sky than to any human form: 'You would think he had three distinct heads of hair – brown at the base, blood-red in the middle and a crown of golden yellow. This hair was settled strikingly into three coils on the cleft at the back of his head. Each long loose-flowing strand hung down in shining splendour over his shoulders, deep-gold and beautiful and fine as a thread of gold. A hundred neat red-gold curls shone darkly on his neck, and his head was covered with a hundred crimson threads matted with gems. He had four dimples in each cheek – yellow, green, crimson and blue – and seven bright pupils, eye-jewels, in each kingly eye. Each foot had seven toes and each hand seven fingers.'

The conclusion to be drawn from the work of Mike Baillie and others is that it is not whether, but when another extreme weather event will darken the Earth. It may come from a passing comet, or from a slumbering supervolcano like the one in California's Yellowstone Park, or it may catch us completely by surprise from somewhere out of left field.

Professor Baillie says: 'There are a lot of things we have no real control over, and from looking at extreme events in the past I would say it really exposes how little we really know about the possible extremes of this planet. We don't know enough about how often we're hit by material from space. We don't know enough about different combinations of cataclysmic eruptions. The real danger is getting a couple of big ones going off around the same time which would have a cumulative effect.

'We've never been in a situation before where we're teed-up with seven billion people on the planet. If you go back 250 years people were quite used to famine and large numbers of people dying, but since the Industrial Revolution it's been "oh no, we can deal with these things". So the expectation is that we can move food around and cope with any downturn, but we haven't had one of these big events in recent times with the result that we don't know if we could cope. The answer, I think, is that we couldn't cope because we all currently import huge amounts of food and in a crisis situation food wouldn't be produced, or it wouldn't be available.

'So I would be pessimistic if such an event took place. It doesn't have to be a Yellowstone. It only has to be something like happened in AD 540 or 1150 BC or 2350 BC. Any of these could be severe enough to cause massive dislocation.'

Keep watching the skies.

IS IT A BIRD?
IS IT A FISH?

The Mysterious Migration
of the Barnacle Goose

I N THE 'PROTEUS' CHAPTER of James Joyce's *Ulysses*, angsty
youth Stephen Dedalus wanders aimlessly on Dublin's
Sandymount Strand reading 'signatures of all things' into the
'seaspawn and seawrack' strewn up by the wind and the waves. He
amuses himself with the thought: 'God becomes man becomes fish
becomes barnacle goose becomes featherbed mountain.'

Given that Joyce's wife's name was Nora Barnacle, it's not hard
to guess the playful association between his lover and a featherbed,
but some of the early Joyce scholars just didn't 'get' the bit where
fish becomes barnacle. One learned gent tied himself in knots
explaining: 'When God, having become man, then becomes fish,
that is no ordinary fish, such as the one Hamlet's beggar ate. It is a
literate Greek fish, an *ichthys*, the result of an artful transubstantiation

by which the flesh (or the fish, if you prefer) is made word.' Enough.

The scholar who penned that erudite piece of codswallop must have been kicking himself when, as must have happened, someone explained that in Irish folklore the Barnacle Goose was considered a fish and that Joyce was having a gentle pop at the Catholic Church which made it an article of faith for hundreds of years.

A thousand years ago and more, the Irish people of the western seaboard would watch huge flocks of these two-tone seabirds head out to sea each spring. We now know they were bound for distant Greenland, but for most Irish people of the time there was no known land to the west, and a superstition grew up to fill the knowledge gap. The Barnacle Geese would disappear for the whole summer, but then return to Irish shores with their numbers replenished by youngsters after what seemed like long months in the open ocean. Never having seen a barnacle goose egg or a gosling, the plain people of Ireland sought other birth channels for these creatures.

The Irish were willing recipients of gifts thrown up by the sea. Big storms might wash up a bounty of kelp for animal feed and fuel, or driftwood for building, or sometimes the makings of an unexpected feast. The Annals of Connacht for AD 1246 report: 'A whale was stranded in Carbury, at Cuil Irra, which brought great relief and joy to the countryside.'

For the dwellers of the western seaboard, the annual return of the Barnacle Geese seemed as much a windfall from the sea as from the skies. The first foreigner to record the mystery of the Barnacle Goose was the Welsh-Norman scribe Giraldus Cambrensis. For a man who never wasted an opportunity to have a go at the supposedly ignorant natives, Giraldus went hook, line and sinker for the yarn that the geese hatched from barnacles attached to driftwood out at sea.

He reported: 'Nature produces [the Barnacle Goose] against Nature in the most extraordinary way. They are like marsh geese but somewhat smaller. They are produced from fir timber tossed along the sea, and are at first like gum. Afterwards they hang down

by their beaks as if they were a seaweed attached to the timber, and are surrounded by shells in order to grow more freely. Having thus in process of time been clothed with a strong coat of feathers, they either fall into the water or fly freely away into the air. They derived their food and growth from the sap of the wood or from the sea, by a secret and most wonderful process of alimentation. I have frequently seen, with my own eyes, more than a thousand of these small bodies of birds, hanging down on the sea-shore from one piece of timber, enclosed in their shells, and already formed. They do not breed and lay eggs like other birds, nor do they ever hatch any eggs, nor do they seem to build nests in any corner of the earth.'

This confusion was a godsend to the monks of medieval Ireland. Abbeys flourished across the land for a thousand years before King Henry VIII shut them down and grabbed their land in the 1540s. For the duration of that millennium the monasteries were the industrial estates of their day and the hub of wealth production. The monks had the best food in the land for their tables, but their favourite meats were forbidden on the many fast days of each Church year and for the forty days of Lent.

Those long weeks of pious abstinence spurred some of the Catholic Church's finest minds to reflect on how Lent might be made less of a Purgatory. They came up with an ingenious and deeply bizarre form of *á la carte* Catholicism to spice up the hotpot of life.

Eight centuries before the Treaty of Rome established the European Economic Community (EU) in 1958, another European economic community was doing brisk business across much of the continent. The Hanseatic League was an alliance of cities which controlled trade right across northern Europe. The League linked the salt mines of the East with the rich fishing grounds of the West and made a fortune supplying all of Europe with salted fish. Even though Ireland was surrounded by water, there was a strong market for imported salted herrings, with the Welsh port of

Porthdinllaen sending over thousands of barrels a year until the start of the twentieth century.

Reports suggest that the salted fish tasted more of salt than fish but it had two massive selling points. The first was that the Catholic Church said you couldn't eat meat on Fridays, fast days, or throughout Lent if you were a Catholic. The second was that the Catholic Church said you had to be a Catholic. For the League, with its virtual monopoly on salted fish, the forty days of fast and abstinence before Easter were like Christmas come early.

But the Universal Church didn't become the Universal Church without having a firm grasp of the fine art of fudge. The Church numbered many of the greatest legal minds of the day in its ranks, who set about the task of finding loopholes with relish.

Rather than condemn everyone to eating salted fish for forty days non-stop, the Church blurred nature's boundaries with a series of sly derogations. By the time the legalistic theologians had finished, beavers, turtles, certain birds and other animals had been reclassified as fish, and therefore kosher to enjoy freshly served during Lent.

Which is where the Barnacle Goose came in. According to one tradition, the Irish churchmen ruled that Barnacle Geese hatched from the vaguely egg-shaped barnacles attached to driftwood, officially making them fish. According to another, they emerged from the buds of the goose barnacle tree, officially making them fruit. After a couple of weeks of salted herring for breakfast, dinner and tea, either must have been greeted as equally plausible.

The monastic island settlement of Skellig and its surrounds on the Kerry mainland enjoyed a special dispensation to eat puffins on fast days. One given reason was that since puffins swim and eat fish, logic dictates that they must be fish. The logic of the Churchmen seems strange today (not to mention strangely convenient), but by the pre-scientific standards of the day it still passed for logic. By the same token the Churchmen were able to make the case that birds with webbed feet were creatures of the sea, and therefore edible as fish, while those without webbing weren't.

TEN WEATHER-RELATED EVENTS
FROM THE ANCIENT IRISH ANNALS

The ancient annals of Ireland were written down by monkish scribes from the beginning of the Christian period in the fifth century AD to the end of the seventeenth century, but they claimed to detail events stretching back to around 3000 BC. The annals started out as a chronology of holy days, but expanded to take in details of political and military events and social upheavals. The monks mentioned episodes of exceptional weather, usually recording them as a sign that The End was nigh, or as written proof of punishments from God on his wicked subjects. The following ten events were thought worthy of mention:

AD 663: Darkness on the Kalends [the first days] of May, at the ninth hour; and in the same summer the sky seemed to be on fire. A pestilence reached Ireland on the Kalends of August. The battle of Luth-feirnn ... and an earthquake in Britain. And Comgan Mac-Ui-Teimhne and Berach Abbot of Bangor (died). Baetan, son of Ua Cormaic, Abbot of Cluain, died. The mortality raged at first in Ireland in Magh Itho of Fothart.

AD 799: There happened great wind, thunder, and lightning, on the day before the festival of Patrick of this year, so that one thousand and ten persons were killed in the territory of Corca Bhaiscinn, and the sea divided the island of Fitha into three parts.

AD 941: A great flood in this year, so that the lower half of Cluain-mic-Nois was swept away by the water.

AD 1328: Much thunder and lightning this year, whereby much of the fruit and produce of all Ireland was ruined, and the corn grew up white and blind.

AD 1329: The cornfields remained unreaped throughout Ireland until after Michaelmas (late September), in consequence of the wet weather.

AD 1339: The cattle and winter grass of Ireland suffered much from frost and snow, which lasted from the end of the first fortnight in winter into spring.

AD 1363: A great wind this year, which wrecked churches and houses and sank many ships and boats.

AD 1373: A very great wind this year, which wrecked many churches.

AD 1471: Showers of hail fell each side of Beltaine [May Day], with lightning and thunders, destroying much blossom and beans and fruits in all parts of Ireland where they fell. One of these showers, in the east, had stones two or three inches long, which made large wounds on the people they struck ... There was another, in the north, which did much damage in Moylburg and at the monastery of Boyle, and a boat could have floated over the floor of the great church of the monks, as we have heard from the folk of that place.

AD 1545: Great dearth in this year, so that sixpence of the old money were given for a cake of bread in Connaught, or six white pence in Meath.

In keeping with their supernatural outlook on the elements, the scribes were prone to spice up their run-of-the-mill weather reports with broadly related sensational items, such as the following:

AD 739: The sea cast ashore a whale in Boirche in the province of Ulster. Every one in the neighbourhood went to see it for its wondrousness. When it was slaughtered, three golden teeth were found in its head, each of which teeth contained fifty ounces. Fiachna, son of Aedh Roin, King of Ulidia, and Eochaidh, son of Breasal, chief of Ui Eathach Iveagh, sent a tooth of them to Beannchair, where it remained for a long time on the altar, to be seen by all in general.

AD 743: Ships with their crews, were plainly seen in the sky this year.

The Irish were not alone in whipping logic through hoops for the sake of a good meal. When Spanish missionaries first encountered South America's giant hamster, the capybara, during the sixteenth century, they wrote to Rome for guidance. Actually, they weren't so much looking for guidance as for the Church to rubber-stamp the admission of yet another reluctant member to the fish family. The missive to Rome read: 'There is an animal here that is scaly but also hairy, and spends time in the water but occasionally comes on land. Can we classify it as a fish?' Despite the fact the capybara isn't remotely scaly, and doesn't have webbed feet, they got the answer they wanted and today most of the 400 tonnes of capybara meat consumed annually is eaten during Lent. In case you're curious, it tastes like pork.

Beavers and seals do have what might be described as webbed feat, and the great seventeenth-century French adventurer in North America, Baron Louis de Lahontan, sarcastically praised the 'doctors' who persuaded a line of Popes to classify beavers, otters and sea-calves as fish for the purposes of the Lenten fast.

At the Fourth Council of the Lateran in 1213 Pope Innocent III slapped a ban on the eating of Barnacle Geese during Lent, decreeing that since they looked like a goose, and walked like a goose, they were most probably geese. As we have seen from the Vatican's verdict on the capybara some 300 years later, the classification of God's creatures remained something of a moveable feast.

Until well into the twentieth century, the Barnacle Goose was known in parts of Ireland as 'the priest's fish'.

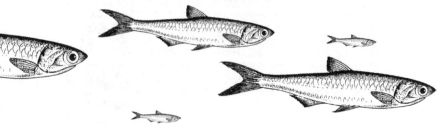

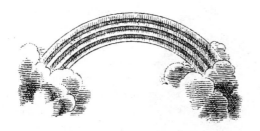

THE ANSWER
IS 42

The Rainbow and the
Crock of Gold

RAINBOWS HAVE BEWITCHED PEOPLE for as long as people have walked the Earth. To the ancient Mesopotamians the rainbow was a goodwill gesture from Mother Earth to the human race. In the *Epic of Gilgamesh*, one of the world's earliest surviving literary works, set down in the cradle of western civilisation, the rainbow was the 'jewelled necklace' of the goddess Ishtar which she lifts into the sky as a pledge that she will never forget the 'great flood' that swept away her human children.

In the Biblical story of Noah's Flood, which clearly recycles the same folk memory of a great deluge as the Gilgamesh tale, the rainbow is again lifted into the sky as a godly token of good faith to mankind. In the Book of Genesis, God promises that he will never again unleash a flood like the one that engulfed the world.

He vows: 'I do set my bow in the cloud, and it shall be for a token of a covenant between Me and the Earth. And it shall come to pass, when I bring a cloud over the Earth, that the bow shall be seen in the cloud. And I will remember my covenant, which is between me and you and every living creature of all flesh; and the waters shall no more become a flood to destroy all flesh. And the bow shall be in the cloud; and I will look upon it, that I may remember the everlasting covenant between God and every living creature of all flesh that is upon the Earth.'

In ancient faraway cultures where the names of Noah or Gilgamesh would have meant nothing, the rainbow was also infused with deep symbolic meaning and was particularly associated with water and snakes. Indeed, in the Dreamtime myth of Australia's isolated Aboriginal people, the Rainbow Serpent was the deity that ruled water.

A rainbow is formed by countless tiny flecks of moisture acting as minuscule mirrors. The beguiling trick of the light is created because the sunlight doesn't bounce directly off the convex front surface of the raindrops, but passes through and reflects off the concave back of the droplet.

As the light passes through the prism of the raindrop, its basic ingredients are separated out into their spectrum of black, red, orange, yellow, green, blue, indigo, violet and white. One of the laws of nature states that a sphere of water always reflects light back at an angle of 42 degrees. The practical effect of this law of optics is that the beauty of a rainbow really is in the eye of the beholder. A rainbow won't appear unless the sunlight catches the droplets at precisely a 42-degree angle to the observer. For the rainbow to be seen, the sun's rays also have to project from behind the observer.

In Irish myth and folklore the rainbow is said to lead to a crock of gold at its base, which may or may not be guarded by bad-tempered

A TO Z
OF IRISH WEATHER

December In Latin '*decem*' means ten, and December was the tenth and last month in the ancient Roman calendar which was discontinued around 713 BC by the traditional (i.e. possibly makey-uppy) second King of Rome Numa Pompilius, who replaced it with a twelve-month calendar.

Deisiol The ancient Irish had a custom of turning sunwise, meaning from right to left, when performing weighty rites and ceremonies. This practice was called *deisiol* from the Irish word *deis* meaning 'the right hand'. The custom transferred from the sun-worshiping pagan Irish to the newly Christianised society, and Saint Patrick is said to have consecrated his church at Armagh by leading the faithful in a solemn procession sunwise around the ground.

Dickens Temperatures plummeted across northern Europe roughly between 1600 and 1850 (the dates are much disputed) in what's called The Little Ice Age. This long period of deep-freeze winters has given us our classic greeting-card image of Yuletide, with crowds of ruddy-cheeked revellers roasting chestnuts, throwing snowballs and making merry on frozen waters. The chief architect of this enduring Christmas card cliché is Charles Dickens (1812–1870) aided and abetted by other artists and writers of similar vintage who would have experienced such scenes in their childhood.

Dog Days John Millington Synge opened his poem 'Queens' with the line: 'Seven dog-days we let pass, naming Queens in Glenmacnass.' The renowned Irish artist Harry Clarke picked up on the theme and used Synge's line as the title of one of his best-known paintings. In ancient Rome the 'dog days' from late July to late August were traditionally the most stifling of the year. The stultifying heat was blamed on the dog star Sirius, and the hot spell was believed to be a time of lurking evil, when the wine turned sour, dogs went mad and fever stalked the land.

leprechauns, depending on where you source your myths. One school of thought maintains that our distant ancestors persuaded themselves that they saw the end of rainbows come to ground at the entrances to burrow mounds and megalithic tombs. These dark places were said to be gateways to the otherworld of the Tuatha Dé Danann and other children of the old gods, who supposedly possessed gold in abundance. Down the course of generations, and under the yoke of Christian orthodoxy, these once impressive children of gods were allegedly belittled into leprechauns.

Or so that particular story goes. There are several other accounts which tell it differently, such as the following which appeared in a volume entitled *The Popular Religion and Folklore of Northern India*. Pointing out that, like the Indians, 'the Celts also had their guardian snake', the Victorian author W. Crooke of the Bengal Civil Service wrote: 'The rainbow is connected to the snake, being the fume of a gigantic serpent blown up from the underground. In Persia it was called the "celestial serpent". It is possibly under the influence of the association of the snake, a treasure guardian, that the English children run to find where the rainbow meets the earth, and expect to find a crock of gold buried at its base.'

Allowing that from a desk in Bengal in the days of Empire, English and Irish children must have seemed all the one, Crooke's suggestion that the Irish rainbow myths stretch way back into the mists of our Indo-European past has much merit. A recurring theme of many end-of-the-rainbow stories is that no one can ever quite lay their hands on the elusive crock of gold. This must be anchored in the realisation that the rainbow itself has no anchor in the ground.

The end of a rainbow can never be pinpointed because the rainbow itself is located in the ether, existing at a different distance and in a different space depending on who is looking at it. A rainbow is not a tangible object. Any observer attempting to creep up on one is on a hiding to nothing. Attempting to observe a rainbow at any angle other than the magic one of 42 degrees will simply make it disappear. Even if one observer can see another observer who appears to be standing 'at the end of' a rainbow, the second viewer will have no sense of being at the rainbow's end. The second viewer will, in fact, be looking at a different manifestation of the same rainbow, but at the same 42-degree angle of the first viewer.

Just as seeing, or not seeing, a rainbow is entirely a matter of your point of view, the delightful spectrum of colours we see is a personal matter. The vivid hues are segregated out by our internal sense of colour perception.

Rainbows are also very much creatures of the hour of day. Since these spectacles can only appear on the opposite side of the sky to the sun, morning rainbows are only seen in the west and evening ones in the east. Because the rainbow can only be seen when the light is reflected at the specific angle of 42 degrees, if the sun is higher than that angle, as it is in the period around noon, no rainbow can ever be seen. Because the sun always appears in the sky to the south of Ireland, every Irish rainbow appears against the backdrop of the northern sky.

WHEN MURPHY SAYS FROST, THEN IT WILL SNOW

The Perils of Forecasting

FROM THE DAWN OF RECORDED TIME, humans have attempted to exert some degree of control over the weather's mood swings by forecasting what it might do next. The people of ancient Egypt lived their lives by the three seasons surrounding the annual flooding of the River Nile. The ritual and stability of that civilisation revolved around the clockwork prediction of the coming flood.

While the ancient Egyptians were chiefly concerned with predicting the flooding caused by heavy rains in mountains far to the south, the Babylonians living to the east in the land we know today as Iraq were interested in watching the skies for more immediate information. By 650 BC, and probably much earlier, the Babylonians were studying cloud formations and other atmospheric signs in an attempt to come up with relatively short-term forecasts. Some three centuries later the Chinese had created a calendar which divided the year into twenty-four festivals, each of which was supposed to chime with a different type of weather.

While certainly not the world's first self-styled weather guru, the Greek philosopher Aristotle did pen the first unified theory of the elements that has come down to us. Written around 340 BC, his *Meteorologica* (*Meteorology*) encompassed observations and theories about all aspects of the weather and climate, from thunder and lightning to floods and earthquakes. Although many of Aristotle's theories of cause and effect were far wide of the mark, his *Meteorologica* remained the Bible of weather forecasters and theorists for some 2,000 years until it was put to the test, and eventually laid to rest, by the appliance of science.

Two centuries or so before Aristotle's ambitious weather treatise, Celtic culture arrived in Ireland – around 500 BC. One of the key feast days on the Celtic calendar was Imbolc, later Christianised to Candlemas, which fell on what is now the first day of February, midway between the winter solstice and the spring equinox. This celebration of the lengthening of the days took different forms across much of Europe, and would involve bonfires, feasting and the watching for weather omens. One element of this weather divination involved keeping vigil to see if badgers and other warmth-loving animals would poke their heads above ground.

The American Groundhog Day ritual of 2 February is a German export of this European-wide spring festival. According to this tradition, if it is overcast when the groundhog (also known as a Land Beaver) emerges from its burrow on this day, spring will come early. However, if it is crisp and sunny, the groundhog will supposedly take fright at its shadow and retreat back into its burrow, indicating that winter will continue for six more weeks.

In Ireland and Celtic Britain Imbolc was believed to be the day the weather hag called the Cailleach gathered up her firewood to see her through the final stretch of winter. Legend had it that if she planned to make the cold last a good while longer, she would conjure up a bright and sunny day so she could gather plenty of fuel for her fire. By this token the people would be gladdened if the day of Imbolc turned out cold and wet, as this would mean the Cailleach was having a short lie in and the winter was almost over.

In Ireland, Imbolc was associated with the goddess Brigid who symbolised the light half of the year. She had the power to guide the community safely from the darkness of winter into the healing light of springtime. When Christianity arrived in Ireland, the goddess Brigid effortlessly switched sides to become Saint Brigid, who assumed many of the same superpowers.

By the time William Shakespeare unveiled his comedy *Twelfth Night* to the public on Candlemas Day 1602, Christianity had not only long absorbed the weather lore of pagan society, but had recently begun a fierce family feud called the Reformation over which popular articles of faith counted as scriptural truth and which could be discounted as pagan-style superstition.

The Twelfth Night of Christmas was a party night that fell on the eve of Old Christmas Day (as calculated by the Julian calendar abandoned in Ireland and Britain in 1752). In Ireland Old Christmas Day became Nollaig na mBan, or Women's Christmas, and was traditionally (or more probably theoretically) the one day of the year when Ireland's housewives could put their feet up while their menfolk waited on them. In Ireland the day is also known as Little Christmas, while abroad it is generally observed as the Feast of the Epiphany. The twelfth day of Christmas was also traditionally the last of the twelve 'ruling days' or 'days of fate' which were said to govern the weather over the coming year. Shortly before Shakespeare penned *Twelfth Night*, his countryman Gervase Markham returned to England after a disastrous stint in Ireland as a captain in the army of Robert Devereux, the Earl of Essex.

As the reign of Queen Elizabeth I came to a close, Markham was an officer in the greatest expeditionary force ever sent to Ireland, numbering some 16,000 men. Instead of subduing the savage Irish, Essex came to a humiliating truce with the Chieftain Hugh O'Neill. Under express orders from his Queen not to return without a victory under his belt, Essex sailed home regardless, to his eventual execution. Markham, meanwhile, returned to his previous career as one of the most prolific writers and great recyclers of his day. He was notorious for repackaging his old material between new covers, to the point that in 1617 his booksellers forced him to give an undertaking that he'd write no more on certain subjects. One of his many books on husbandry (the care and cultivation of crops and livestock) was *The English Husband*. In it he wrote of the 'ruling days' of Christmas that: 'What the weather shall be on the sixth and twentieth day of December, the like shall it be in the month of January; what it shall be on the seventh and twentieth, the like shall it be the following February; and so on until the Twelfth Day, each day's weather foreshadowing a month of the year.'

Gervase Markham died in 1637. Exactly 200 years later in 1837 a Corkman living in England named Patrick Murphy published a prediction that 20 January 1838 would be exceptionally cold. After years of authoring unheeded works on weather, climate, atmospherics, electricity, magnetism and the movements of the heavens, Murphy became an overnight sensation thanks to his single-sentence entry for 20 January which said: 'Fair, and probably the lowest degree in winter temperature.'

That line appeared in his *Weather Almanack On Scientific Principles, Showing The State of the Weather For Every Day of the Year 1838*. It was one of many almanacs published in that period purporting to predict the far-distant weather, but Murphy got lucky. On the day in question the temperature in London plummeted to either

14 °C or 20 °C below zero, depending on who you believe. The Thames froze over so solidly that people walked from one side to the other. Thousands were said to have died amongst the poor, while the better-off enjoyed beer and skittles on the ice. On the River Medway in Maidstone, a sheep was roasted on a spit on the frozen water. It was reported that a novel type of cricket was played on a lake near Tunbridge Wells where all the players wore skates.

The cold snap was dubbed Murphy's Winter by a suitably impressed public, and copies of his *Almanack* began to fly off the shelves. It ran to over forty reprints and made Murphy a large fortune, estimated at £3,000. As the year of 1838 wore on, however, much of the sheen wore off Murphy's accomplishment as readers studied his predictions with great expectations. At the end of the year, the sceptics pronounced the forecaster a chancer, pointing out that Murphy's predictions had come close enough on 188 days, but had been way off the mark on 197 days.

The Irishman continued to produce his annual weather almanac until a couple of years before his death in 1847, but to ever diminishing sales returns. His great windfall from the 1838 edition also fell victim to his shortcomings as a clairvoyant, and he quickly blew it speculating on grain prices. As his fortunes waned, Murphy was lampooned in a play entitled *The Man With The Weather Eye*, in which a learned gentleman mistakes a lumpen potato seller for the famous meteorologist.

Even *The Times* of London wasn't beneath taking a dig at the discredited forecaster, joining in with the following verse:

When Murphy says 'frost' then it will snow,
The wind's fast asleep when he tells us 'twill blow,
For his rain we get sunshine, for high we have low.
Yet he swears he's infallible — weather or no!

GHOST SHIPS AND CRYSTAL PILLARS

Brendan the Navigator
and the Arctic Mirage

THE FANTASTIC VOYAGE OF SAINT BRENDAN the Navigator in the early Middle Ages was one of the great adventure stories that was much admired and retold across Medieval Europe. One of the earliest preserved versions is the Dutch *De Reis Van Sint Brandaen* which was written in the twelfth century, and which itself is thought to have been sourced from a lost text in the High German script. While Brendan's epic tale contains many far-fetched elements, scholars agree that the original *Voyage of Saint Brendan the Navigator* is informed by the distilled seafaring knowledge of generations of Irish navigators.

The Irish mastery of the Sea of Perpetual Gloom was down to the superb construction and design of the currach, a boat made with sewn–up skins and a well-proportioned keel that could survive the roughest swells of the North Atlantic. The currach appeared on the west coast of Ireland around 500 BC, and by the time of Saint Brendan's voyage 1,000 years later the Irish had used these exceptional craft to explore far beyond their own shores.

The best-known story of Brendan's adventures in the northern ocean tells of how he celebrated Easter Sunday Mass on what he thought was a small island, but which turned out to be a snoozing whale. Tall tales like this shouldn't distract from the genuine maritime achievements of the seafaring Irish who came before and after the celebrated saint.

In AD 825 the Irish monk Dicuel, writing in his Frankish monastery, told of how much earlier Irish hermits had discovered and settled the Faroe Islands situated far to the west of Norway. The Faroes are located almost exactly halfway between Norway and Iceland, although it seems no one knew that when the first explorer monks arrived on the faraway islands, as no one knew that Iceland even existed.

Saint Columba was a contemporary of Brendan's and the text entitled *The Life of Saint Columba* states that Iceland was found at the end of an unlimited ocean. So are we to believe that the monks set out westwards from the Faroes on a hunch, or might they have followed nature's signposts?

In their 1993 book *Histoire Du Nouvelle Monde II*, Carmen Bernand and Serge Gruzinscki suggested that the Irish seamen may have been led (or misled) by an Arctic mirage. They wrote: 'From time to time, particularly in summer, which in the early medieval era was the normal sailing season, when air rests on a colder surface and the observed image may be optically displaced in a vertical direction from its true location … (this) permits the occasional

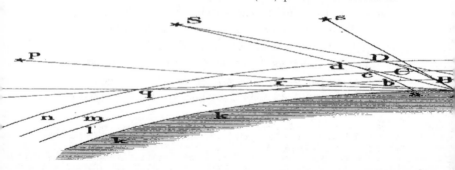

sighting of objects such as islands and mountains situated far beyond the normal distance of the horizon. The rising sun creates the conditions most propitious for this phenomenon.'

In the Arctic, people can sometimes see and hear things that they cannot see or hear elsewhere on Earth. Microscopic ice crystals suspended in the air change how light and sound travel over distances. Layers of hot and cold air refract, or bend, light rays. Light bounces off the surfaces of clouds, water and ice to create optical illusions.

The optical phenomenon of the Arctic mirage, also known as a superior mirage, occurs because of special atmospheric conditions that bend light. A superior mirage occurs when an image of an object appears above the actual object. The illusion is created by a weather condition known as a temperature inversion, where cold air lies close to the ground with warmer air above it. Since cold air is denser than warm air, it bends light towards the eyes of someone standing on the ground, changing how a distant object appears.

Superior mirages are common at polar latitudes, especially over large sheets of ice that have a uniform low temperature. When the conditions are right, they can appear at lower latitudes and it is very possible that the Irish seafarers in the region of the Faroes would have witnessed these impressive spectacles. A far-distant coastline might appear much taller, and therefore much nearer, than it really was. By the same token, a coastline which would normally lie over the horizon, and therefore be out of sight, might be projected to appear above the horizon. Again, the optical illusion would be convincing enough to persuade the Irish sailors that they were looking at dry land within reach, when they were actually seeing an image projected onto thin air.

This illusion is caused by the way that light waves bend in keeping with the curvature of the planet. By following this curvature, light rays can travel from far beyond the horizon. This was documented for the first time in 1596 by the Dutch explorer Willem Barents who was attempting to find a Northeast Passage which would allow trading vessels to sail above and around Siberia and down to the Far Eastern lands of spice and silk.

The Dutchman's ambitious expedition became trapped in sea ice in the Arctic Ocean and had to dig in for a polar winter siege. They were astonished when, two weeks earlier than their calculations said it should happen, the long spell of midwinter darkness was ended by a warped sun appearing in the sky. This miracle remained a mystery for more than three centuries until scientists came to the conclusion that Barents' wobbly sun was actually below the horizon, but that its rays had bent with the curve of the Earth to project it into the sky. Today the effect is known as the Novaya Zemlya mirage.

The superior mirage provides a plausible explanation for the legend of the Flying Dutchman, a ghost ship cursed never again to see its home, but to sail the seven seas for all eternity. In his 1910 book *Round-about Rambles*, Frank R. Stockton wrote the following in a chapter entitled 'When We Must Not Believe Our Eyes':

> Once a vessel was sailing over a northern ocean in the midst of the short, Arctic summer. The sun was hot, the air was still, and a group of sailors lying lazily upon the deck were almost asleep, when an exclamation of fear from one of them made them all spring to their feet. The one who had uttered the cry pointed into the air at a little distance, and there the awe-stricken sailors saw a large ship, with all sails set, gliding over what seemed to be a placid ocean, for beneath the ship was the reflection of it.
>
> The news soon spread through the vessel that a phantom-ship with a ghostly crew was

sailing in the air over a phantom ocean, and that it was a bad omen, and meant that not one of them should ever see land again. The captain was told the wonderful tale, and coming on deck, he explained to the sailors that this strange appearance was caused by the reflection of some ship that was sailing on the water below this image, but at such a distance they could not see it. There were certain conditions of the atmosphere, he said, when the sun's rays could form a perfect picture in the air of objects on the Earth, like the images one sees in glass or water, but they were not generally upright, as in the case of this ship, but reversed – turned bottom upwards. This appearance in the air is called a mirage. He told a sailor to go up to the foretop and look beyond the phantom ship.

The man obeyed, and reported that he could see on the water, below the ship in the air, one precisely like it. Just then another ship was seen in the air, only this one was a steamship, and was bottom-upwards, as the captain had said these mirages generally appeared. Soon after, the steamship itself came in sight. The sailors were now convinced, and never afterwards believed in phantom ships.

For a more recent incident, we can turn to the late author and broadcaster Ludovic Kennedy. In his book *Pursuit: The Chase and Sinking of the 'Bismarck'* he described a mysterious Second World War encounter off the Denmark Strait between Iceland and Greenland. Pursued by the British Navy, the German battleship *Bismarck* fled into a sea mist, only to reappear in an instant steaming towards its pursuers at high speed. The British cruisers took evasive action, but then the crews watched flabbergasted as the German vessel flickered and disintegrated. The British radar confirmed that the oncoming *Bismarck* had been an optical illusion.

If Arctic mirages didn't signpost the way to Iceland for the Irish monks of the Faroes, there was an unmissable highway in the sky pointing them in that direction. Every spring and summer huge numbers of birds fly north from Ireland, Scotland and the rest of Europe to make the most of the explosion of plants, small mammals, fish and especially insects during the brief but fecund Arctic summer. The easterly winds that carried the birds towards Iceland would have summoned the Irish along the same route.

Dicuel wrote in his chronicle that in January 795 a band of hermits left Ireland, arriving in Iceland on the first day of February. The French-based monk reported that they made their home

A TO Z
OF IRISH WEATHER

Earthquakes Ireland sits in one of the planet's most earthquake-free zones, but they are not unknown. In June 2012 an earthquake of a magnitude of four on the Richter Scale was recorded 60km west of Belmullet, County Mayo. Conspiracy theorists have claimed the disruption was caused by work on the Corrib Gas Field, but the fact is that other quakes have been recorded in the vicinity of Ireland which don't fit the theory.

Easter The celebration of Easter is a moveable feast, meaning that it does not follow the civil calendar but is linked to the cycles of the moon and sun. Disagreements over how to calculate the precise date of Easter led to bitter squabbles between rival Church factions across Ireland and Britain in the Early Middle Ages.

East Wind Moisture is always evaporating from our skin into the air. If the air is dry, more evaporates, making our skin feel cooler. This is one reason why, in Ireland, a dry wind from the east feels much colder than a moisturised wind from the Atlantic. We perceive the west wind as warmer because more moisture stays in our skin.

Equinox An equinox occurs twice a year (around 20 March and 22 September) when the Earth's axis is tilted so that the sun appears to fly directly above this planet's equator. The practical effect is that on the equinox, night and day are of roughly equal length on every part of the planet.

there until the start of August. Dicuel said that members of that expedition of thirty years earlier gave him first-hand accounts of experiencing the so-called Land of the Midnight Sun.

He reported: 'Not only at the very summer solstice, but also at a late hour of the days immediately before and after, the setting sun hides himself, as it were, behind a low hillock, so that there comes no darkness for even the shortest space of time, but a man can, as in the sunshine, do any work he pleases, were it so much as catching lice on his shirt.'

The timing of the Irish hermits' voyage rings true with what has come down to us of their navigation techniques. Five hundred years before the magnetic and the gyroscopic compasses made open-sea voyages a much less risky business, the Irish sailors used the sun and the stars for guidance. Today it is woven into the folk tales of the Faroe Islanders that the Irish set out for Iceland early in the year to avoid the fogs and the bright nights from May to early August that made the map in the heavens hard to see.

Dicuel was a propagandist for the Christian Church, and for the heroic escapades of his fellow Irishmen. It is telling, then, that in writing of the Irish visit to Iceland in 795 he didn't make the slightest suggestion that this expedition had discovered the Land of Fire or that the voyage was in any way out of the ordinary. It is clear from his account, and from others, that the new arrivals of 795 were welcomed into an existing encampment.

The great nineteenth-century Icelandic scholar Finn Magnussen conceded: 'It would be a great mistake to hold that Iceland was first discovered by the Norwegians, for the ancient Icelandic historiographer Are Frode … clearly states that the newcomers [the Vikings] found in the eastern parts of the island certain Christians, whom they called Papas, and who were Irishmen, as was apparent from the books written in Irish, which, among other things, they left behind at the time of their departure.' The Icelandic records mention three principal settlements of the Papas in the east of the country, and one in the west, where the next stop would have been Greenland.

There is no proof that the Irish made what would have counted as a transatlantic crossing to Greenland. However, that possibility is raised by several sources, like Dicuel's mention that the Irish knew of a 'frozen sea' a day's sail to the north of Iceland. This strongly suggests that the Irish sailed around Iceland, travelling north until they encountered this frozen sea. This circumnavigation of the island would have taken them through the seaway known today as the Denmark Strait where the phantom *Bismarck* seemed to switch instantly from retreat to attack.

It is entirely plausible that an Arctic mirage would have lifted the shores of distant Greenland onto the horizon, making them seem much nearer than they actually were. Several passages in the *Voyage of Saint Brendan* suggest a knowledge of things stretching further to the west. For instance, Brendan's Pillar of Crystal sounds very much like an iceberg, which could have been found only in an area of ocean near Greenland. His terrifying tusked monster fits the bill of the North Atlantic walrus. His short, swarthy natives could have been the Inuit of Greenland, and his foggy banks boiling with fish sound remarkably like the once-teeming coast of Newfoundland.

WAVES OF MUTILATION

Flood Myths and Tsunamis

A T THE BEGINNING OF THIS MILLENNIUM two geologists made the disturbing prediction that sooner or later a massive landslide on the Canary island of La Palma will start a so-called mega tsunami which will devastate the coasts of southern Ireland and Britain, while wreaking havoc across the North Atlantic from Africa to the Americas.

Other scientists slammed as 'scaremongering' the claim that a wave taller than Dublin's Spire will hit Ireland at 500 mph and flatten everything in its path for five miles inland. Happily, Ireland's past experience of tidal waves suggests that the chances of a tearaway torrent raging far inland are remote in the extreme.

The geological fact is that giant waves are most common in the

Pacific Ocean, which is circled with shifting cracks in the Earth's crust called subduction zones. Apart from a couple of unstable areas in the Caribbean and off the coast of Argentina, the Atlantic is blessed with what geologists call 'passive margins'.

So while the Atlantic weather that batters Ireland's shorelines is often regarded as rougher than that of the Pacific (named for its more placid nature), giant killer waves are a very rare occurrence. Strangely enough, though, the ancient tales which make up the foundation myths of the Irish people tell of several great floods and giant waves.

The Book of Invasions, for instance, tells of the arrival in Ireland of the goddess Cessair, the granddaughter of Noah, along with several superhero companions. Cessair eventually drowns in a Great Flood that sweeps her father Bith to the north of the country and Ladra to the east, where the storytellers claim he dies 'of an excess of women'. Fintan survives the deluge by turning himself into a salmon and setting up his home in the centre of the country, from where he divides Ireland into its five ancient provinces.

A later group of supernaturally gifted arrivals were the fairy folk called the Tuatha Dé Danann who were closely associated with waves. The tide at Glandore Harbour in County Cork is to this day known as Tonn Chliodhna or Cliodhna's Wave after one of the Banshee queens of the newcomers. The Celtic text known as *Dinnseanchas* (meaning *Lore of Places*) says: 'She was a foreigner from Fairyland, who, coming to Ireland, was drowned while sleeping on the strand at the harbour of Glandore in South Cork. In this harbour the sea, at certain times, utters a very peculiar, deep, hollow, and melancholy roar, among the caverns of the cliffs, which was formerly believed to foretell the death of a king of the south of Ireland. This surge has been from time immemorial called Tonn-Cleena, Cleena's wave.'

Other significant great waves recorded in the foundation myths occur in the north of the country. One was said to have swept up the mouth of the River Bann and carried off Tuagh, the beautiful daughter of the High King. Another was the Wave of Ruair, the Tonn Ruarí, which was said to rise in Dundrum Bay in County Donegal and to have come to the aid of Ulster when King Connor Mac Nessa beat upon his shield in times of need. The Wave of Assaroe, which is said to have brought great carnage to the Donegal countryside around Ballyshannon, is invoked in the ancient text *The Destruction of Da Derga's Hostel*.

One or some, or none, of these great wave tales may have been folk memories of a giant tsunami which fanned out from the coast of Norway around 6,100 BC, when most of the small scattering of people in Ireland seem to have been settled along the northern coastline. The huge wave was triggered by underwater subsidence in the Storegga Slides, an area of the continental shelf known in Norwegian as 'the Great Edge'. The tsunami generated by the landslide activity is thought to have swept right over some of the Shetland Islands, and to have laid deposits of debris about two feet thick far inland from the Scottish coastline.

Gathering evidence of past great waves is a relatively new scientific pursuit, but the harder the experts look, the more they appear to find. In 2009 and 2010 Dr Sander Sheffers of Australia's Southern Cross University co-authored new papers on coastal silt and boulder deposits in Galway Bay and on the Aran Islands. His findings added weight to other recent studies which suggest that a number of tsunamis hit the west of Ireland in the last millennium, with one concluding:'From all field observations (boulder ridges in very high positions, apart from the modern coastline, with extremely large clasts) and datings (clustering in particular for medieval times and the 18th century) we only can conclude, that several tsunamis of different energy and a more regional impact occurred during the last millennia. The tsunami history of the northern Atlantic Ocean is just in the beginning of its deciphering.'

A TO Z
OF IRISH WEATHER

Fahrenheit Measuring the temperature of water from freezing point to boiling point on a scale from 32 °F to 212 °F, the system devised by Daniel Fahrenheit (1686–1736) was the European standard until eventually replaced by Celsius in the twentieth century. The United States is the only major country to retain it as its official scale.

February The Roman month Februarius takes its name from the Latin word '*februum*', meaning 'purification'. The rains of spring were thought to wash away the winter gloom and purify the land.

Freckles and Red Hair Freckles are clusters of the pigment melanin usually caused by the exposure of pale skin to the sun. They are particularly associated with red hair. In 1931 Harvard University's Department of Anthropology began a study of the Irish population of a type that would not be dreamed of today, owing to the fact that it had its basis in discredited racist thinking. It was eventually published under the un-PC title *Structure, Head Form and Pigmentation of Adult Male Irish*. The report concluded that some 10 per cent of the population had red or auburn hair (a figure since borne out as accurate), and that the greatest concentration of redheads was in the north of the country, with the lowest numbers found in Wexford and Waterford. More recent and more scientific studies have found that red hair occurs naturally in only 1 to 2 per cent of the global human population, rising to 4 per cent amongst Europeans. Red hair does appear to be favoured in places where sunlight is at a premium, like Ireland. It is associated with fair skin colour, reflecting low concentrations of eumelanin (the pigment most abundant in black skin). This lower melanin concentration confers the advantage that enough Vitamin D can be produced under low light conditions to keep the body healthy.

Front Ireland owes its rainy, unsettled weather in large part to the fact that it lies on the boundary where the cold easterly Arctic air mass comes into contact with the warmer westerly air mass that lies further south. The shifting boundary, called the Polar Front, can stray southwards as far as Iberia and the Mediterranean during the winter, and can move as far north as the Shetlands during the summer. Unfortunately for the Irish people, the unsettled Polar Front spends a great deal of each year passing directly over Ireland while yo-yoing one way or the other.

In 2012 the distinguished archaeologist Alan R. Hayden wrote a piece for the *Journal of the Kerry Archaeological & Historical Society* suggesting that the southwestern coast has endured several giant waves that have lived on in the folk memory. By cross-referencing the folklore with the hard evidence, Hayden concluded that the cluster of Valentia, Beginish and Church islands may bear the scars of sudden impacts from the sea.

One folk tale collected in the locality early in the twentieth century provided a striking name for a stretch of ancient road running from Dolus Head on the mainland through Valentia and on to 'Glen Smoil at the foot of the Skelligs'. While Hayden dismissed the notion of the land route going to the foot of the far-out Skelligs as 'fanciful', he was intrigued by the name of that road, Bóthar na Scairte, or Road of the Cataclysm. Hayden unearthed two legends concerning a local hero by the name of Clusagh, who was either a great seafarer or a great hurler. The story goes that one fine day a hurling contest was taking place in Loch Cé, a now submerged area between Beginish, Church Island and Dolus Head. The 'multitude of gentry' that came to watch the game fled in panic when 'the sky darkened and a terrible wave fifty feet high' came rushing in upon them. All bar Clusagh managed to escape to high ground, but the land remained flooded forever after, and Skellig and Valentia became islands.

We know for certain that tidal waves have hit the south and west of Ireland in historical times. The Portuguese capital, Lisbon, was famously rocked by an earthquake and a sixty-foot tsunami in 1755, and was hit again in 1761. The epicentre of the quakes was off the Portuguese coast, but the disturbance caused disruption along the coast of Ireland and as far north as Scotland.

Writing of the 1755 Portuguese wave, the Scots geologist Charles Lyell noted:

> The agitation of lakes, rivers, and springs in Great Britain were remarkable. At Loch Lomond, in Scotland, for example, the water, without the least apparent cause,

rose against its banks, and then subsided below its usual level. The greatest perpendicular height of this swell was two feet four inches. It is said that the movement of this earthquake was undulatory, and that it travelled at the rate of twenty miles a minute. A great wave swept over the coast of Spain, and is said to have been sixty feet high in Cadiz. At Tangier in Africa it rose and fell eighteen times on the coast. At Funchal in Madeira it rose full fifteen feet perpendicular above high-water mark, although the tide, which ebbs and flows there seven feet, was then at half ebb. Besides entering the city and committing great havoc, it overflowed other seaports in the island. At Kinsale in Ireland, a body of water rushed into the harbour, whirled round several vessels and poured into the marketplace.

Other reports said that the first tsunami arrived on the Irish coast four hours after reaching Portugal, and that it was followed by several more terrifying waves over the next seven hours. Part of Galway's famous Spanish Arch was destroyed, while the churning waters gouged out deep chasms along the County Clare coastline at Killomoran, Caherglissane, Gort and Kinvara. A castle at Coranroe on the north coast of Clare was flattened.

The wave(s) struck on 1 November 1755, which was observed by the Catholic natives as All Saints' Day, and one folk tale tells of a miller who was said to have brought the disaster on his townland of Kinvara. He reportedly offered up a prayer to both God and the Devil to help him to grind his corn, as there was not enough water to power the mill. For this act of blasphemy, he was rewarded with more water than his mill could cope with. The mill, so the story goes, was destroyed and the miller drowned for his sins.

WHEN WEATHER CHANGED HISTORY PART 1

Edward de Bruce's Irish Invasion, 1315

EDWARD WAS THE YOUNGER BROTHER of Robert de Bruce, aka King Robert the First of Scotland. The brothers shared a vision of founding what they called a Pan-Gaelic Greater Scotia. This would be an enlarged version of the sixth-century Irish sea kingdom that embraced parts of Ulster and western Scotland, but with its axis of power now tilted firmly towards Scotland. The de Bruces determined to forge a family fiefdom incorporating Scotland and Ireland, with Robert as King of Scotland and Edward as High King of Ireland.

With Edward at his side, Robert secured his rule over Scotland with victory over England's King Edward II at the Battle of Bannockburn in 1314. In May of the following year the younger de Bruce arrived in Ireland with a force of some 6,000 men. His aim was to open a second front against the English and to have himself installed as Ireland's first High King since that position became vacant in 1186. Edward was no stranger to Ireland. It's believed he was fostered as a child to an aristocratic Gaelic family in Ulster, possibly the O'Neills. He could also trace his lineage back to Ireland's most revered High King, Brian Boru, who had ruled three centuries earlier.

Gaelic chieftains flocked to join Edward's cause, and this grand Gaelic alliance cut a swathe through the country, winning skirmish after skirmish and scoring a resounding victory over the English at the Battle of Kells. It seemed that nothing could stop de Bruce from taking over the entire island when the weather dealt his ambition a cruel and decisive blow in the winter of 1316/7.

The Great Famine of 1315–1317, as it became known, was sparked by a spate of foul, cold, wet weather in the spring of 1315 that destroyed crops in the fields from Russia to Ireland. Food production in 1316 was equally disastrous, with a seed yield as low as 2:1, meaning that for every seed planted, just one grew for food needs, and one to plant for the next year's crop. Food prices in Ireland soared. Salt, the chief way to cure and preserve meat, was almost impossible to obtain because the process of evaporation used to extract it couldn't take place in the damp weather, so the price of it, too, skyrocketed.

Famine killed millions across the continent as far south as Italy. In Ireland and elsewhere the catastrophe was taken as a punishment from God. In England, the unknown poet who penned *The Evil Times of Edward II* wrote: 'When God saw that the world was so over proud / He sent a dearth on earth, and made it full hard / A bushel of wheat was at four shillings more / Of which men might have had a quarter before.'

What should have been Edward de Bruce's triumphant

procession through Ireland became bogged down in the climactic catastrophe which reduced large parts of Europe to anarchy and even cannibalism. It's believed that the German folk tale of Hansel and Gretel, which tells of a brother and sister abandoned to the clutches of a hungry witch, dates from this Great Famine.

Unable to feed his large army in famine-stricken Ireland, Edward's hitherto unstoppable campaign ran into the ground. His big brother, Robert, who had joined him in a final push for victory, now retreated home to deal with the effects of the crisis in Scotland, promising to send fresh aid as soon as possible. Edward and his remaining forces settled in for months of stalemate, as the Anglo-Irish barons ranged against him were crippled by the same inability to put an army in the field.

Improved weather in 1318 allowed both sides to resume hostilities, with serious consequences for Edward. The self-proclaimed new High King of Ireland was killed in the Scots/Irish defeat at the Battle of Faughart. His body was hacked into take-away portions which were put on display in several Irish towns, while his head was sent as a cheering memento to England's King Edward II.

Had the sun shone on Edward de Bruce's Irish venture, the history of Ireland might have been very different, with English power nipped in the bud and Irish ways and Irish laws restored, even if paying temporary lip-service to a Scots dynasty, and even if the respite was only temporary. In the event, Edward's campaign marked the last concerted effort that stood any real chance of ousting English colonial rule and making Ireland a Gaelic nation once again. Irish chieftains would win some famous battles over the centuries that followed, but the war against the foreigner was effectively lost.

RAIN AND SLIME

The Curse of the Pishogue

THE *LONDON JOURNAL* OF 9 JANUARY 1909 carried the following report filed by an English tourist recently returned from Ireland. It was reproduced in a 1930 book entitled *The Origins of Popular Superstitions and Customs*. Introducing the story, the book's author, T. Sharper Knowlson, remarked that it illustrated how 'the evil eye is said to have a dread terror for the more ignorant Erse (Irish) population'.

The report read: 'Some months ago I was on a visit to some friends in the south of Ireland, and one morning when seated at breakfast a servant rushed into the room screaming hysterically that the dairymaid has just found pishogue upon the dairy floor. Pishogue is a white, yellowish fungus made at the dead of night, after a solemn incantation of the devil, according to a secret rite which has been handed down from generation to generation.

'My host, a "big" landlord, sprang to his feet and, followed by his wife and myself, ran hastily out of the house into the trim, cool dairy where, upon the posts of the door, I saw the daubs of pishogue. My host knocked it off quickly with a stick, and then, turning angrily to the weeping dairymaid, told her it was nothing at all. But the next minute he informed me under his breath that he might expect bad luck with his dairy.

'That very evening when his twenty splendid milch cows were driven into their stalls to be milked, a cry of consternation went up from the lips of the milkers; they were absolutely dry. And for months they remained so, while a tenant who lived close to the demesne – an absolutely drunken impecunious rascal – was noticed to give up his weekly attendance at Mass, in spite of which irreligious conduct his miserable dairy stock suddenly took the appearance of healthy, well-fed cattle, and every one knew he was the man who had put pishogue on his master and robbed him of his good.'

The meaning of the term 'pishogue' is elusive. It was a catch-all word which could refer to the casting of an evil spell, or the evidence of that evil spell, or a slice of bad luck. The term is often found in relation to superstitions surrounding cattle and milking, and especially in cases where cows stop giving milk, such as the one described in *The London Journal*.

It seems certain that the 'pishogue' that brought the big landlord dashing from his breakfast table was the jelly-like fungus *Tremella mesenterica* which has a variety of folk names including Yellow Brain, Golden Jelly, Yellow Rambler and Witches' Butter. A yellowish-orange in colour, it is slimy to the touch and a good fall of rain brings it out into a gooey, garish bloom. The damp, rotting wood of old barns makes a perfect habitat for Witches' Butter, which dries into a thin, barely noticeable, crust, only to plump up again like some giant pus-filled pimple when the rain returns.

WEATHER AND WITCHCRAFT

Hags, Harvests And Hatreds

ONE OF THE GREATEST MULTITASKING DEITIES of the ancient Irish was the Cailleach, a wild woman who wore a veil to signify her unfathomable mystery. She had powers over the birds of the sky and the beasts of the fields, she could take the form of rivers and mountains, but she was perhaps most strongly connected to the whims and moods and terrifying tantrums of the weather.

The Cailleach was venerated in place names and shrines throughout Ireland, and across Scotland where she was known as The Old Wife (or Old Hag) of Thunder. She was to be feared and respected because, as the goddess of winter, she had the power of life and death over communities that were never more than a few square meals away from starvation.

She was 'the daughter of the little sun' who grew more powerful as the days grew shorter and the weakened sun flew lower in the sky. She brandished a slachdan (wand of power) with which she could conduct the weather. Wherever the Cailleach threw her slachdan nothing would grow. As the sun regained its strength the Cailleach would lose hers, before she was finally overthrown at the spring equinox in March, which was the ancient New Year's Day.

The Cailleach appears to have first cousins in the Irish Banshee and the Welsh Hag of the Mist, both of whom could be heard wailing on the wind when someone was about to die. The more distant relations of these supernatural beings were the wise women who, until relatively recent times, would provide their communities with herbal remedies, spells and potions.

When it came to converting from paganism to Christianity in the fifth and sixth centuries, the Irish had taken the path of least resistance by keeping on virtually all the old deities, customs and beliefs, while giving them new names and enough of a Christian gloss for respectability.

So when Pope Innocent VIII issued a Papal Bull in 1484 declaring war on witches, the Irish-speaking population paid as little attention to it as they had when his predecessor, Sergius III, declared in 906 that witches didn't exist at all. The main plank of Pope Innocent's crusade against witches in 1484 was that through their evil powers over pestilence and the weather, they 'have blasted the produce of the Earth, the grapes of the vine, the fruits of the trees'.

The Pope's calls for a European witch-hunt fell largely on deaf ears at the time it was issued, but his Bull took on a new lease of life in the period roughly between 1520 and 1750. This age of witch-hunting corresponds strongly with the religious strife that followed Martin Luther's rebellion against Rome in 1517, but many climatologists also believe it was fuelled by a severe chill that gripped Europe.

Experts don't agree on when the so-called Little Ice Age started and ended, but they do agree that the temperature drop across

Europe was at its sharpest from the late 1500s to around 1850. Many researchers over the years have linked the stressed growing conditions of this period with stressed communities. From Ireland to Russia, every society was a subsistence society. One failed harvest could mean hunger and great hardship, but two or more in succession could mean famine, sickness and death. When basic resources were in short supply, neighbours could become rivals for the little there was.

In such times of cold, scarcity and hunger, people also looked for someone to blame. Witches, with their supposed powers over the weather, became obvious targets. In 2004 the Harvard scholar Emily Oster published a paper showing a remarkable match between the sharpest temperature drops of the period and the persecution of witches. Others had noticed the correspondence, but Oster was the first to map temperature against trial records, decade by decade. She established that one of the steepest single temperature drops, around 1560, coincided with a sudden resurgence in European witch trials after a lull of seven decades.

Across Europe the craze for hunting and burning witches was at its most intense in parts which had most enthusiastically embraced the new Protestantism. The persecutions, prosecutions and executions were at their worst in Germany, where the official death toll was 8,188, but where the historian Ronald Hutton believes as many as 26,000 may have actually been killed. In proportion to their much smaller populations, the slaughter in the hard-line Calvinist heartlands of Switzerland and Scotland was just as great.

Meanwhile, the horrors of Europe's witch craze passed almost unnoticed in Gaelic Ireland where the old-time religion continued to accommodate the mystic hags with their spells and potions. Herdsmen might occasionally rant and rave that a so-called Butter Witch had spitefully syphoned off the milk from their cows, leaving them dry, but it was the Protestant planters who introduced the idea of the Satanic Witch that had to be hunted down and killed.

Of the tiny handful of prosecutions for witchcraft that took place in Ireland during the period, all were conducted by the planters against their own kind.

Left to go about their business more or less undisturbed, the mystic women of Catholic Gaelic Ireland would surely have applauded the memorable observation of Father Ted Crilly: 'That's the great thing about Catholicism. It's so vague ...'

IRISH WEATHER PROVERBS
– ANIMALS

If the seal comes up to the seashore, bad weather lies ahead.

Rain is due when a herd of cows lie down together in the middle of a field, reluctant to rise.

February kills the sheep.

Rain is coming when the dog dozes off during the day.

When the fox barks at night a dry spell is promised.

Rain is coming if sheep start playing rough with one another.

Good weather clouds are like sheep's wool – bad weather clouds are like goat's hair.

Cattle sense autumn, and smelling the harvest they begin to trespass.

March is the month of the brindled (tiger-striped) cow.

Sheep gather on the top of a hill awaiting fine days.

If the hedgehog is not seen before a moonlit night in May, soft foggy days are due.

If the dog's stomach rumbles the weather will change.

If a man drinks sow's milk he can see the wind for evermore.

Bad weather is coming when the dog eats grass.

WHEN WEATHER CHANGED HISTORY PART 2

The Spanish Armada and the Protestant Wind, 1588

'Jehovah blew with His winds, and they were scattered.' That phrase, printed in Latin, appeared on commemorative English medals minted shortly after the defeat of the Spanish Armada in 1588. The gales which scattered the invasion fleet from Catholic Spain became known as the Protestant Wind. Exactly 100 years later the English would credit a second Protestant Wind with helping to topple a revived Catholic monarchy in England and replace it with a Protestant regime that had almost universal popular backing. (See pp.100–104)

But back to 1588. In May of that year Spain's stoutly Catholic King Philip II assembled a fleet of some 130 ships in Lisbon with orders to pick up 30,000 troops waiting in the Spanish-controlled Netherlands. The Armada's commander, the Duke of Medina Sidonia, was charged with landing a total of 55,000 soldiers on English soil and overthrowing the Protestant Queen Elizabeth I. Philip had not only gathered an army that was absolutely massive for its time, but he had the blessing of the Pope and a good claim to the English throne. Philip had been the King of England and Ireland for four years (1554–58) until the death of his wife, Queen Mary I, thirty years earlier. The county of Offaly was named King's County in 1556 in his honour. He dismissed Mary's half-sister Elizabeth as a heretic and a usurper.

The so-called Protestant Wind first blew up in Philip's face when heavy weather slowed the Armada's progress from Portugal to the English Channel to a crawl. Four galleys and one galleon were forced to turn back early on, and by the time the Spanish fleet came within range of their Dutch-based invasion force, six weeks of hanging around had allowed disease to get a grip and the numbers waiting to board the ferry barges had almost halved.

The liberation of Catholic Ireland from Protestant English rule was barely an afterthought in Philip's grand plan, although it would be a nice bonus prize in the war of religions consuming Europe six decades after Martin Luther had kick-started the Reformation by nailing his Ninety-Five Theses to the door of Wittenberg's Castle Church. Ireland, however, would become the graveyard for Philip's grand plan.

The 200-strong English fleet out-thought and outfought the Spanish off Gravelines in northern France, and forced the Armada to abandon its plans to pick up what was left of its invasion force in the Low Countries. The English then chased the Spanish north to the Firth of Forth on the east coast of Scotland, leaving Medina Sidonia with no option but to take the long way home around the north of Scotland and Ireland.

As the Spanish entered the choppy North Atlantic, they attempted to give the coast of Ireland a wide berth, but measuring longitude at that time was still an inexact science bordering on guesswork. While the Spanish mistakenly believed they were making headway westwards towards open sea, the Gulf Stream was actually pushing them northwards and eastwards. The catastrophic upshot was that when they finally turned south, they were much closer to the rocky Irish coast than they'd bargained for.

Trouble piled upon trouble when a series of strong westerly gales blew up, driving the ships leeward towards the jagged shores. In order to speed away from the pursuing English fleet, many of the Spanish ships had cut loose their heavy anchors, which left them now unable to fix themselves in secure shelter. Europe was entering what is called the Little Ice Age and 1588 was marked by particularly fierce storms in the North Atlantic, which may have been associated with a build-up of polar ice off Greenland.

Between seventeen and twenty-four Spanish ships came to grief on a stretch of Irish coastline from Antrim in the northeast of the country to Kerry in the southwest, with the loss of some 5,000 men, half of them drowned and the other half killed or captured. On a single day, 21 September, fourteen ships were sent to the bottom by hurricane force winds. With English troops thin on the ground in the west of Ireland, some commanders opted to kill their captives rather than risk their escape or rescue. Survivors who expected help from friendly natives were worse than disappointed: they were slaughtered by the Gaels. Shortly after the mopping-up operation, Queen Elizabeth was told that only around 100 survivors remained in Ireland.

Of the 130 ships that had set out from Lisbon, only 67 made it home to Iberia carrying 10,000 survivors, many of whom were diseased and dying after long, torrid weeks at sea. It is said that when

King Philip II learned the full extent of the disaster, he groaned: 'I sent the Armada against men, not God's winds and waves.'

The Spanish losses off the coast of Ireland greatly exceeded those in the skirmishes with the English navy. As the grieving Spanish put it themselves, the flower of Spain's nobility was cut down in Irish waters. The mauling suffered by the Armada marked a tipping point in European power politics. From then on, a newly self-confident English state was on the rise while the seeming invincibility of the formidable Spanish Empire suffered a damaging blow.

TOFFEE FOR THE NORTHERN CLIMATE

Seasonal Sweets and Ices

IN AN ERA WHERE WE THINK NOTHING of picking fresh strawberries, apples or grapes from the supermarket shelf in midwinter, it is perhaps hard to imagine a time when most fruit and veg only came around at a certain time of year. What is perhaps even harder to imagine is that less than a century ago even sugary manufactured sweets had their seasons.

Although chocolate back then was a special treat at any time of year, it was marketed well into the 1950s as a nutritional snack to keep out the chill of winter. Cough drops, clove drops, aniseed balls, mints and anything with a vaguely medicinal taste was flogged as a sure-fire way of keeping the airways unblocked.

By the 1930s toffee had begun challenging these old reliables for the position of the nation's number one winter sweet. As one confectioner wrote in an Irish trade journal: 'Peppermint is always largely consumed as a flavour during the winter months, and toffee flavoured with it was taken into favour some years ago. Somehow the unique combination of so many good things – milk, butter, sugar – by an ingenious process, has taken hold of the public taste so that it will not willingly dispense with its toffee during the winter months. There is this to be said for both toffee and chocolate, and perhaps most of all for toffee, that in this northern climate they both have a distinctively warming effect upon the human circulation, and seem to make the cold air itself more intense. They are both largely nourishing.'

The most highly seasonal sweet of them all, however, was ice cream. A century ago, before the availability of electric refrigeration, ice cream in Ireland was both very exclusive and very expensive. This was largely to do with the fact that it was made from blocks of ice cut by hand out of the mountain lakes of Norway and shipped south to London, Dublin and other parts of the British Isles.

By the 1920s artificial factory-made ice was becoming widespread, but ice cream remained a rare and pricey luxury made and sold by some sixty Italian families, mostly in Dublin and Cork. A number of good summers in the 1920s helped turn ice cream from a premium product at a premium price into an affordable treat for the masses. The change came about as small dairies across Ireland amalgamated into bigger ones. After a good grass-growing season in the spring and early summer of a given year, the dairies would have a supply of milk much larger than they required for the pint bottles on the doorsteps and the manufacture of chocolate and butter.

The bright idea of the dairy owners in the 1920s was to pump the surplus milk into the manufacture of ice cream. In 1933 one of the new supersized dairies, Hughes Brothers, launched their pint brick of HB ice cream. The brick was shrewdly targeted at the new suburbs that were beginning to sprout up around the capital, and by the close of the decade it had become an affordable Sunday treat for

A to Z
of Irish Weather

Glaciers The landscape of Ireland today, including rivers, lakes, mountains and valleys, was formed by the advance and retreat of a vast ice sheet during the last Ice Age. The so-called Irish Sea Glacier, which covered most of Ireland, Wales, Scotland and the Isle of Man, was probably at its most extensive some 22,000 years ago and began retreating around 13,000 years ago.

Global Warming Since the beginning of the twentieth century the mean temperature of the Earth's surface has increased by about 0.8 °C. Two thirds of this increase has occurred since 1980. Greenhouse gases produced by human activities have been blamed by non-mad scientists. The warming greenhouse effect was first proposed by the French physicist Joseph Fourier as early as 1824.

Gregorian Calendar In 45 BC Rome adopted the new Julian calendar of Julius Caesar which tidied up the ancient Roman calendar which had fallen badly out of synch with the seasons. But Caesar's method of calculation was also flawed, and by 1582 the Julian calendar was fully ten days out of whack with the solar year. So that year Pope Gregory XIII issued a Papal Bull instructing that 5 October would become 15 October, losing the ten unwanted days and putting the calendar back on track. Europe's Catholic states made the changeover with minimum delay, but many Protestant regimes rejected it despite its obvious merits. In a letter of 1584 English officials in Dublin complained to London that the leading Irish chieftain Hugh O'Neill and his Catholic Gaelic allies had been 'solemnizing their New Easter of the Pope' in defiance of their Protestant overlords. Britain, and by extension Ireland, did not adopt the Gregorian calendar for almost two centuries more – in 1752 – by which time it was necessary to subtract eleven days to correct the difference.

Going A term to describe the condition of a racecourse in horse racing. Depending on the weather, the going can be hard, firm, good, soft, yielding or heavy.

many families. In fact, as more and more people were moved out of the crumbling inner-city tenements to the new suburbs, having ice cream after Sunday dinner gained a snob value with the new class of upwardly mobile estate dwellers.

Years later, the availability of ice cream, milk and chocolate in the nation's shops still depended to an extent on the whims of the weather. The uncommonly hot and dry spring and summer of 1949 stunted the growth of grass which led to low milk yields which in turn caused a shortage of milk and ice cream in Dublin city and county that summer. Press reports held that two huge new holiday camps in Mosney and Skerries north of Dublin were to blame for swiping milk and ice cream from the lips of ordinary decent dwellers of the capital.

Four years later, in May 1953, *The Sunday Press* reported that thanks to a combination of mild weather and a big spurt in the growth of grass, all of the country's chocolate factory workers had been put on overtime. According to the newspaper: 'Dublin's chocolate factories are taking on more and more hands in order to keep working right around the clock to cope with the greatest supply of surplus milk that ever flowed into Dublin. One city supplier said: "I've never seen anything like it."'

WHEN WEATHER
CHANGED
HISTORY
PART 3

The Battle of Kinsale, 1601

I N 1598 THE SPANISH MINISTER Diego Brochero de Anaya
wrote to his king, Philip III, in praise of the Irish Legion which
had been a part of the Spanish Army since the 1580s. He urged
that: 'Every year Your Highness should order to recruit in Ireland
some Irish soldiers, who are people tough and strong, and nor the
cold weather or bad food could kill them easily as they would
with the Spanish, as in their island, which is much colder than
this one, they are almost naked, they sleep on the floor and eat

oatbread, meat and water, without drinking any wine.' The Irish brigade fought Spain's war against the breakaway Protestants of the Netherlands until 1600, when it was disbanded due to heavy losses wrought by combat and sickness.

A year later, in 1601, veterans of the Irish Legion touched back down on home ground under the Spanish flag. Thirteen years after much of the Spanish Armada had been dashed on Irish shores by stormy weather, the Spanish dispatched a smaller fleet to Ireland carrying an invasion force of some 6,000 troops. Once again, bad weather took a hand in the course of Irish history. Thrown off course by fierce storms, nine of the ships turned back for Spain, taking with them 2,000 of the most battle-hardened troops and most of the gunpowder.

When the remaining ships dropped anchor in October 1601 with 4,000 soldiers, they landed at the wrong end of the country. They took the County Cork town of Kinsale with little fuss, but their main Gaelic allies Hugh O'Neill and Hugh O'Donnell were holed up in their Ulster strongholds, almost 300 miles away. The Spanish were quickly besieged by a 7,000-strong English force led by Lord Mountjoy, with reinforcements bringing that number up to 12,000.

The Spanish commander Juan del Aguila sent word north for O'Neill and O'Donnell to march to his aid. The chieftains hesitated for weeks as the autumn turned into a particularly severe winter, but with the Spanish running low on supplies and morale, the country's two most powerful Gaelic leaders reluctantly set out with a combined force of 5,000 infantry and 700 cavalry. The two Irish armies travelled separately in the knowledge that even minimal food rations would be hard to secure in the depths of winter.

The Victorian poet Aubrey de Vere glorified the miserable trek in his poem *The March To Kinsale*, which opened with the rousing lines …

O'er many a river bridged with ice,
O'er many a vale with snow-drifts dumb,
Past quaking fen and precipice
The princes of the North are come.
Lo! Those are they who year by year
Roll'd back the tide of England's war;
Rejoice Kinsale, thy help is near,
That wondrous winter march is o'er.

There was a brief, wondrous respite for the exhausted Irish troops when a freezing snap turned the mushy bogs into rock-hard thoroughfares, allowing the heavily laden Ulster armies to cross in

A TO Z
OF IRISH WEATHER

Hail Hailstones are the product of the updrafts associated with thunderstorm clouds. Like rain, hailstones begin life as water droplets carried in the cloud. Carried vertically upwards they become supercooled as the temperature drops. As the forming hail rises, it passes through layers of varying humidity and temperature, at some points absorbing moisture in a process called 'wet growth'. Eventually its weight can no longer be supported by the updraft and it begins to fall.

Harvest Moon/Hunter's Moon The harvest moon is the full moon that occurs closest to the autumnal (September) equinox. The latest possible date for a harvest moon is the second week of October. Tradition has it that farmers could work late into the evening bringing in their crops by the light of the harvest moon. The first full moon following the harvest moon is the hunter's moon, which was said to provide the ideal light for hunting migrating birds returning in great flocks from their summer Arctic feeding and breeding grounds.

safety, skipping around English obstructions. One of the English commanders, Sir George Carew, later lauded the Irish feat as 'the greatest march with carriage [baggage] that hath been heard of'. O'Neill and O'Donnell reached Kinsale to find the besieging English army was in a sorry state, desperately short on food and brought low by thousands of deaths and desertions inflicted by the hostile elements.

Unfortunately for the newly arrived Irish, the Spaniards holed up in Kinsale weren't in much better shape and, with his men reduced to a diet of rusks and water, del Aguila urged the Gaelic chieftains to attack. O'Neill was unsure. He felt that the now surrounded English were cold, wet and starving and very close

Hay Fever Hay fever is an allergic inflammation of the nasal airways (allergic rhinitis) caused by pollen, principally during the haying season when plants are cut and stacked for winter feed. It is possible to suffer from hay fever at any time of the year. While the cause and effect varies with individuals and place, the chief culprit is the minuscule pollen of wind-pollenated plants which shoot out vast clouds of spores in a blunderbuss effort to reproduce. The pollens of plants pollenated by insects are generally too heavy to stay aloft in the air, but Ireland has an abundance of trees which produce the airborne pollens that spark hay fever. Scientists believe birch is the worst offender, followed in no particular order by pine, cedar, hazel, horse chestnut, poplar and willow. Strong winds and heavy rain are the allies of sufferers, as the former dilutes the concentration of spores in the atmosphere, while the latter washes them to the ground.

Hedging Described by the distinguished Irish meteorologist Brendan McWilliams as 'the honourable ploy of hedging', this is the practice whereby weather forecasters 'hedge' their bets by predicting every type of weather for the coming period in the knowledge that they're almost bound to get enough of it vaguely right to muddle through.

Humidity This is the amount of moisture in the air, in the form of invisible water vapour. The higher the humidity, the less effective sweating is to cool the body, since it becomes harder to evaporate moisture from the body into air that is already moist.

to breaking point. His military successes had come as a master of guerrilla hit-and-run tactics, and he knew the English troops were better drilled in the art of the full-on frontal attack.

But O'Neill's partner in the venture, Hugh O'Donnell, took the Spanish line that an all-out attack was the best strategy. He reminded O'Neill that the men they had force-marched from Ulster were tired, hungry, and falling ill at an alarming rate. It was now deepest December and the biting cold and pouring rain were sapping away the strength and morale of men who had only the most basic shelter against the elements.

The dismal weather was probably the decisive factor that swayed O'Neill to go against his better instincts and attack. Legend has it that the Irish plans were betrayed by a spy, but according to the historian Edward Alfred D'Alton, writing in 1912, the weather added a cruel betrayal of its own. D'Alton wrote of the attack launched on Christmas Eve 1601: 'The night was dark and stormy, the guides lost their way; and when they arrived at the English trenches, weary, exhausted, and dispirited, the morning of the 24th had dawned, and they found the English quite ready, horses saddled, men standing to arms.'

Defeat at the Battle of Kinsale spelled the end for the last strongholds of autonomous Gaelic Ireland. O'Neill and the new O'Donnell fled the country for Spain in 1607 with a motley entourage of lesser chieftains and camp followers. The so-called Flight of the Earls would leave a power vacuum which would be swiftly filled by a strengthened English administration and the plantations of large numbers of Scots and English colonists.

A Tax on Heat and Light

Hearth Tax, Window Tax
and the Half-Door

THE PICTURE-POSTCARD WHITEWASHED COTTAGE has become such a defining image of Ireland that it's easy to think of it as being as much a part of the timeless landscape as the hills and rocks and streams. In fact, the stone cottage is a relatively recent newcomer to the nation's building stock, dating back perhaps only 500 years or even less. One school of thought is that cottages began to spring up as native peasants copied the building techniques used to erect the English style houses of their social betters.

The first function of any domestic building is to keep the elements at bay, and cottage builders followed the ancient rule that the front door faced south wherever possible. Some were built with a door front and back, so that fresh air and light could enter no matter what direction the wind blew from.

The source of heat was the hearth, which was usually made from stone and set at the centre of the dwelling with a bedroom behind it to make for snug sleeping.

The hearth was the social heart of the cottage. Everything revolved around it, from cooking to storytelling. The all-important fire was never allowed to go out, and ashes were strewn over it at night to keep the embers alive for revival in the morning.

Poor farm families measured their meagre wealth in livestock, and it was a common practice to keep the animals very close to hand, safe from the elements and from rustlers. Dwellings that housed humans and beasts under the same roof were known as byre cottages. The Old English word 'byre' means 'storm' or 'strong wind'. Byre cottages were normally built with sloping floors. The animals were penned at the lower end of the slope so that their waste could drain off, away from the human living space. In a variation on the byre theme, some dwellings were built on two storeys, with the family living in a loft space above their livestock. This provided the animals with shelter, while the body heat they generated travelled upwards to warm their owners above.

However good their design and construction, the dwellings of the toiling Irish peasant class were invariably dark and smelly. The bad smell was worsened after dark if rushes soaked in fish oil were used to provide extra light. Candles were beyond the household budget of most families.

The better-heeled classes of society had to endure heating and lighting challenges of a different kind. As two of the basic necessities for human life, heat and light have been targeted as cash cows by kings, dictators and democratic governments since records began.

In 1784 a brick tax was imposed on Ireland by Westminster to fund the losing war to keep the American colonies British. This levy of four shillings on every thousand bricks hit the manufacturers, who, as ever, passed on the hike to the householder. When the manufacturers briefly got around the tax by simply making fewer but bigger bricks, the government set a maximum legal brick size. Timber and weatherboarding didn't provide the same protection against the cold and damp, but they were much cheaper and they enjoyed a big revival in the following decades until the brick tax was scrapped in 1850 as a block to enterprise.

The Hearth Tax, or Chimney Tax, or Chimney Money, was supposed to be a progressive fuel tax that would hit the rich hardest. The rich, needless to day, didn't take this lying down. Under duress, they paid the annual levy on the multiple fireplaces and stoves in their mansions. However, while they were supposed to pay the chimney tax on the labourers' cottages, rectories and other properties on their lands, of which they were the landlords, they used their clout to tweak the law to make their tenants liable.

Hearth tax was universally hated. Since it was impossible to tell how many hearths there might be in a big house from the outside, an army of private collectors was licensed by the authorities. These Chimney Men had the legal power to go through the homes of the middle and upper classes room by room, and their visits aroused huge anger and resentment. There are many stories of the lady of the house throwing a tantrum and barricading herself into a room or a wing, in order to keep the fireplace snoops at bay.

Householders caught trying to get around the hearth tax by blocking up their chimneys were punished with a demand for double the original amount. The Dutch prince William of Orange was invited to become King of England in 1688 by that country's Protestant propertied class. Part of his deal with his rich backers was that hearth tax, which hit them hardest, would be ditched. This deal came five years after four people died in a fire caused by a baker trying to evade the tax by concealing his oven. As a punishment on the Irish who had defied William at the Battle of the Boyne, the hearth tax was kept on in Ireland to raise funds and as a database for keeping a close eye on potential troublemakers.

Having abolished hearth tax as an election promise, William of Orange quickly replaced it with Window Tax. This was decried as a reprehensible 'tax on light', and some scholars persist (wrongly, I'm certain) with the notion that window tax gave us the saying 'daylight robbery'.

Still burdened with hearth tax, Ireland was spared window tax until the aftermath of the disastrous 1798 Rebellion. It was then imposed to fund Britain's war against Revolutionary France, but also to payroll 130,000 occupying troops stationed here to keep the natives submissive.

While the hovels of the poor were exempt from this tax on light, the far grander homes of the rich and famous could be easily assessed for it without the fuss of sending inspectors knocking down the front door. Now a bank, the old Irish Parliament building on Dublin's College Green was built with hollows where the windows were intended to go once the tax was scrapped. Three centuries on, the window tax is long gone but the windows were never installed. For the thirty years before window tax was abolished in 1851, buildings with seven or fewer windows were exempted. Accordingly, many substantial houses surviving from the time have exactly seven windows.

Today the rustic half-door is considered a quaint and charming feature of old Irish buildings. But the half-door, aside from keeping children in and livestock out, was another means of beating the window tax by letting in plenty of light while, for revenue purposes, not being a window.

WHEN WEATHER
CHANGED
HISTORY
PART 4

Cork's Revenge on Cromwell

IN THE SUMMER OF 1649 Oliver Cromwell landed in Dublin at the head of a Roundhead army sent to wipe out Irish resistance to the newborn English Commonwealth. The New Model Army of Parliament had emerged victorious in the English Civil War against the Royalist forces of King Charles I, who had been beheaded the previous January. Parliament now needed to consolidate its regime change.

The Civil War in England had allowed Ireland's Catholic nobles to re-establish a broad degree of independence from London, and

Cromwell identified this pro-Royalist armed camp as a mortal threat to the infant Commonwealth. He arrived in Ireland a man of fifty years, in rude good health and still close to his prime. He left stricken with a sickness that would incapacitate him throughout his reign as the Uncrowned King of England, and would ultimately kill him before he could firmly bed-in the roots of the Republican revolution.

A TO Z
OF IRISH WEATHER

Indian Summer Imported from North America, this has become standard currency in Ireland to describe a last fling of unseasonably pleasant weather before the full onset of winter.

Irishman Joke The earliest recorded Thick Paddy joke appears in the greatest joke book of the Middle Ages and it revolves around the weather. The Florentine scholar and all-round Renaissance Man Gian Francesco Poggio Bracciolini (1380–1459) served in high office under seven Popes, but found international fame as the author/compiler of the best-selling *Liber Facetiarum*, a compendium of gags.

Be warned in advance that this joke isn't funny any more. It tells of an Irish sea captain whose craft was caught in a terrible storm. The tempest was so fierce that he feared all on board would perish, so he prayed to the Virgin Mary for help. The Irishman promised the Virgin Mary that if she saved his ship from the storm he would build her an offering of a huge votive candle the size of the mast of his ship. Dismayed at this, one of his crew warned the Irishman that he mustn't make such a false promise, since there wasn't enough wax in all of England to construct such a candle. The Captain shushed the speaker, telling him: 'Be quiet! Let me promise what I want to the Mother of God – for when we're all saved she'll be as happy with a penny candle.'

The punchline, in case you missed it, is that the Thick Paddy thought he could put one over on the Mother of God. This Irishman joke was clearly considered sufficiently side-splitting at the time to warrant inclusion in the top joke book of the age. Funny peculiar, that.

 The brutality of Cromwell's Irish campaign is a matter of record, and the massacres at the fortified towns of Drogheda and Wexford are numbered amongst the darkest days of Irish history. In contrast, he secured the surrender of Cork by persuading its Protestant Royalist defenders to switch sides and hand over the city. But that easy victory would come at a heavy personal price.

The name 'Cork' translates as 'marsh' and it was while campaigning in the boggy terrain of Cork that Cromwell first succumbed to a fever which would return time and again throughout what should have been the most triumphant decade of his life. When he had revived sufficiently to record the episode that had laid him low, he wrote that he had been 'very sick' and 'crazy in my health'.

Medics now believe that the general was stricken with malaria in Cork. The man who took pride in sharing so much with his troops now shared in the 'ague' or 'country sickness' which felled hundreds of them campaigning in the Munster marshes where a combination of warm weather and marshland created ideal conditions for mosquitoes to breed. Today, Ireland has eighteen species of mosquito, four of which belong to the malaria-carrying species *Anopheles*. In the seventeenth century, malaria was relatively common in damp, mild, lowland regions of Europe, and Cork fitted the bill perfectly.

Cromwell left Ireland in the spring of 1650 carrying a permanent souvenir of his unhappy visit. Just months later, in September, after securing a landmark victory against Scots Royalists at the Battle of Dunbar, he confessed to his wife that his Irish illness had knocked some of the life out of him. He wrote: 'I assure thee, I grow an old man, and feel infirmities of age marvellously stealing upon me.' The fever struck again in the New Year of 1651, and Cromwell took to his sickbed in Edinburgh for some four months.

Worried that the figurehead of their revolution was at death's door, Parliament sent doctors from England to treat him. On the road to recovery, but still weak and sickly, he wrote: 'I thought I should have died of this fit of sickness.' It had been, he said, 'so violent that indeed my nature was not able to bear the weight thereof'.

The fever caught in Ireland took its toll on Cromwell's immune system and lowered his defences against attacks of dysentery, abscesses, boils and other infections that plagued his final years. In the summer of 1658 he was laid low for a full month with yet another bout of his recurring malarial fever, but by now the 59-year-old was literally on his last legs. In September of that year the illness returned for the last time and carried him away.

The Uncrowned King of England was succeeded by his son Richard, who lacked his father's authority and charisma. Bereft of a power base in either the parliament or the army, Richard was forced to step down sheepishly in May 1659. Oliver Cromwell's illness had denied him the chance to bed-in the English Republic, and in 1660 the rudderless state invited Charles II back from exile to restore the monarchy. The new regime also restored the celebration of Christmas, which had been cancelled by Cromwell and his fellow Puritans in 1647 as a Popish plot to make merry with no basis in scripture.

A MAN FOR ALL SEASONS

Ireland's First Weatherman, William Molyneux (1656–98), and the Birth of the RDS

ORN INTO A WELL-TO-DO Anglican family in Dublin in 1656, William Molyneux crammed an impressive number of firsts into his brief lifespan of just forty-two years. He was the first to translate into English the hugely influential work *Meditations* by René Descartes, the so-called Father of Modern Philosophy. He was the first to publish an authoritative English-language treatise exploring the new scientific field of optics. He was the founder of the Dublin Philosophical Society which would eventually become the Royal Dublin Society (RDS). He was also Ireland's first scientific weather observer.

At the age of thirty Molyneux published a slim volume entitled *Sciothericum Telescopicum* in which he outlined his design for a new instrument he called his telescopic sundial. He believed that his sundial clock would revolutionise astronomy and make seafaring safer, but his faith wasn't shared by many and the invention was quietly shelved.

In the preface to *Sciothericum Telescopicum* Molyneux set out the chief aims and methods of the new Dublin Philosophy Society. These included Navigation, which was vital to the success of the Royal Navy, and which required further research into the Earth's magnetic fields and the tides. For his own part, he was taking regular measurements of the high and low tides in Dublin Bay and he'd invented a hydroscope for measuring the amount of moisture held in plants, cloths and other objects. He also earmarked Meterology, or the study of weather, as a key field of future science requiring investigation. Throughout his adult life he encouraged others to follow in his footsteps by designing and inventing new scientific devices which might prove useful. The future, he saw, was with technology. Unhappily, much of the existing technology wasn't fit for purpose. He bemoaned: 'I am living in a kingdom barren of all things, but especially of ingenious artificers. I am wholly destitute of instruments on which I can rely.'

In 1683 Molyneux became the first Secretary of the new Dublin Philosophical Society to promote the advancement of science. The following year he began his *Weather Register* which included barometric data on air pressure and has a place in posterity as the first scientific weather record kept in Ireland. In June 1684 he impressed the members of the Dublin Society with a detailed paper, given the unwieldy title *The Observations of the Weather for the Month of May, with the Winds and the Heights of the Mercury in the Baroscope*.

Although Molyneux's *Weather Register* was discontinued after two years, it remains a milestone in the story of weather observing in these islands. While he continued to study various aspects of the climate, from eclipses to bog drainage, he kept other irons in the fire right to the end of his life. He formulated a mental experiment known as Molyneux's Problem. It is still debated today and was incorporated into the arguments of the great philosopher John Locke which influenced the leaders of the American Revolution.

Shortly before his death in 1698, Molyneux published another paper which would stoke the debate about independence from Britain, both in Ireland and the American colonies. While the last thing Molyneux wanted was a break with England, his call for fair treatment in (deep breath) *The Case of Ireland's Being Bound by Acts of Parliament in England, Stated*, led to calls for his imprisonment as a subversive.

Almost a century after his death, with American independence just achieved, Henry Grattan declared the Irish Parliament free and independent from Westminster with the rousing words: 'Spirit of Swift, spirit of Molyneux, your genius has prevailed! Ireland is now a nation!'

The short-lived Grattan's Parliament came to grief in graft, chaos and the imposition of the Act of Union with Britain in 1801. The short-lived William Molyneux, on the other hand, is remembered as Ireland's first proper weatherman, and much more to boot.

WHEN WEATHER CHANGED HISTORY PART 5

The Protestant Wind – Part 2 and The Battle of the Boyne, 1688–90

S HORTLY AFTER THE SPANISH ARMADA of 1588 was repelled by the English navy and then battered onto the jagged Irish coastline, a greatly relieved Queen Elizabeth I sanctioned the minting of medals to commemorate how a so-called Protestant Wind had scattered her Catholic enemies. Precisely 100 years later a Dutch prince, William of Orange, gathered a new invasion force in the Netherlands and awaited a Protestant Wind to blow from the east and carry his ships to England.

A century earlier, the Armada had planned to pick up an army of 30,000 men in the Spanish-controlled Low Countries to invade England and overthrow the Protestant regime of Elizabeth with a Catholic one run by Spain and Rome. Now, the tables were turned. The Protestant Dutch had thrown out their Spanish overlords and eagerly answered a call from England's Protestant ascendancy who had seen their power and influence fade since Catholic King James II had come to the throne three years earlier.

James had ascended the throne with promises to maintain the Protestant status quo, but it quickly became clear that he intended to secure not just freedom of worship for Catholics, and for marginalised Protestant dissenters, but he also wanted to allow Catholics back into public office.

By the summer of 1688 England's most powerful churchmen, landowners and businessmen were on the verge of mutiny. The mutinous mood spilled over into a fully fledged coup after James' wife gave birth to a healthy son in June, whose arrival dashed all hopes that the king would die childless and the crown would pass to his Protestant daughter, Mary.

A group of nobles petitioned Mary's husband, William of Orange, to topple James from the throne. William, who also happened to be a nephew of James, didn't need to be asked twice, but throughout September and October he was confined to port by a 'Popish Wind' blowing from the west.

According to the eminent Victorian poet, politician and historian Thomas Macaulay, 'the gales which at times blew obstinately from the west prevented the Prince's armament from sailing, and also brought fresh Irish regiments from Dublin to Chester, where they were bitterly cursed and reviled by the common people. The weather, it is said, "is Popish". Crowds stood on Cheapside gazing intently at the weathercock on the graceful steeple of Bow Church, praying for a "Protestant Wind".'

William's 'Protestant Wind' finally arrived in November 1688. The strong easterly wind not only carried William's fleet swiftly across the English

Channel, but it also prevented James from reinforcing his wavering army with any more troops from Ireland. William's landing was greeted with anti-Catholic rioting and mass desertions from the king's army. After a couple of skirmishes, James fled to France and his Dutch son-in-law was installed as King William III. William's supporters hailed it as a relatively bloodless Glorious Revolution.

Years later, one of William's key sponsors, Bishop Gilbert Burnet, wrote in his memoirs: 'A foolish ballad was made at that time,

A TO Z
OF IRISH WEATHER

Jack Frost In the days before double glazing and central heating, people would often awake on a freezing morning to find serrated, spindly figures of ice etched on the inside of their window panes. These patterns can often still be seen on car windscreens on icy mornings. These doodles were attributed in folklore to Jack Frost, although in fact they were the natural outcome of humidity in the room (or vehicle) forming as ice crystals on the freezing glass pane. Jocul Frosti (Icicle Frost) was an elfin creature with an artistic streak from Viking lore. In Mother Russia the same character is known as Father Frost, and in the German Fatherland as Mother Frost.

January The first month of the year is named after Janus, the two-faced Roman god who looked to both the future and the past. Janus was a god of many things, including time and movement, but his job description could be summarised in the broadest sense as 'the God of Beginnings'.

Jet Stream High in the upper atmosphere above Ireland, oscillating somewhere between 7km to 10km (23–39,000 feet) the northern polar jet stream zips around the globe from west to east moving at much greater speeds than the winds below. The meandering air current can push and pull cyclonic storm systems in the atmosphere beneath, and in recent decades meteorologists have tracked the path of the jet stream as a useful tool of forecasting. The aviation industry has been

treating the Papists, and chiefly the Irish, in a ridiculous manner …
that made an impression on the army that cannot be well imagined
by those that saw it not. The people both in the city and the country
were singing it perpetually.' The song, which is thought to feature
pidgin Gaelic, was called 'Lilliburlero' and contained the lyric: 'Oh,
but why does he stay behind? / Lilliburlero Buellenala / Ho, by my
soul, 'tis a Protestant Wind.'

The extended final act of the Glorious Revolution was played

exploiting the jet stream since 1952 when a Pan Am flight from Tokyo
to Honolulu hitched a ride on the current at an altitude of 7,600 metres
and radically shortened the flying time from 18 to 11.5 hours. Similarly,
apart from saving fuel, using the jet stream can slice an hour or more
off inward flights to Ireland from North America compared to the flight
out.

July The Roman month of Quintilis was renamed July in honour of the
assassinated dictator Julius Caesar who, according to tradition, was
born in the month of Quintilis in 100 BC.

Julian Calendar In 45 BC Julius Caesar
replaced the ancient Roman calendar with
a new one of 365 days divided into twelve
months with a February leap day added every
four years. Caesar's reformed calendar was a marked
improvement on what had gone before, which was
badly out of kilter with the solar year, but it wasn't quite
perfect. Because of a discrepancy of mere minutes
each year, the Julian calendar drifted from its moorings
in the real world. By 1582, when Pope Gregory XIII
introduced his own replacement calendar, the spring
equinox was falling on 11 March according to the
Julian calendar, when in the physical world it fell ten days later.

June The sixth month of the year in both the Julian and Gregorian
calendars, June is, according to one tradition, named after the Roman
goddess Juno, who took a special interest in women's rights and
marriage. June was a particularly popular month for weddings in
Ancient Rome, and remains so in the northern hemisphere today.

out in Ireland between 1689 and 1691. It ended with military victory for William, permanent exile for James, and with the Flight of the Wild Geese which saw some 25,000 men, women and children flee Ireland for the continent with their leader, Patrick Sarsfield. The turning point came at the Battle of the Boyne in 1690, an encounter much delayed by the Irish weather. William had sent Marshal Frederick Schomberg to Ireland in the summer of 1689, but the German general had become bogged down by the foul Irish climate. For weeks his army camped near Dundalk face-to-face with an enemy force, both eager to do battle, but the weather was so wet and miserable that Schomberg eventually gave orders to withdraw. His army spent a freezing, damp winter living off scraps of food while thousands of their number died of disease.

Furious with Schomberg's inactivity, William decided to take the future of his tenuous kingdom into his own hands and landed at Carrickfergus with 300 ships and an assembled force of 36,000 men. It was midsummer, and with the weather firming up the ground for the type of pitched battle then in vogue, he defeated James at the Boyne on 1 July 1690. Britain adopted a new Gregorian calendar in 1752 which moved the date of the annual Boyne commemoration to 12 July.

James fled the battlefield for the safety of France, earning himself the Irish nickname 'Séamus an Chaca' ('James the Shit'). William punished his rebellious Irish subjects with even harsher penal laws, and settled down to a fairly undisturbed rule until his death in 1702.

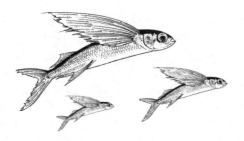

HERRINGS WERE FOUND SIX MILES INLAND

The Night of the Big Wind, 1839

O N THE FIRST DAY OF 1909, the first Old Age Pensions
Act became law, entitling anyone over the age of seventy
in the United Kingdom of Britain and Ireland to claim
a yearly payment of £13 from the state. The move was not as
progressive as its backers liked to claim, since very few people lived
to see seventy.

To draw the new pension a person had to 'be of good character'.
Ineligible persons included those in receipt of Poor Relief,
institutionalised 'lunatics', and anyone who had been in prison
during the previous ten years. Individuals who had been convicted
of drunkenness could be excluded, as could able-bodied folk who
exhibited a 'habitual failure to work'.

The weather was to be a significant factor in the administration
of the new pension. Many people in Ireland and Britain who

believed, or at least hoped, they qualified had no documentary proof that they were born in or before 1839. Local committees were established to vet each claim.

During the first three months of 1909, 261,668 applications were made in Ireland. Proportionally, this was far greater than the take-up in England, Scotland or Wales. *The Times* of London poked fun at the enthusiastic uptake in Ireland, commenting: 'The Chancellor of the Exchequer provoked the merriment of the House of Commons in this connection a month ago by giving the estimated number of persons over seventy years of age in the United Kingdom. He showed that, on the basis of the figures quoted by him, the percentage of the persons claiming old age pensions in the population over seventy years of age in Ireland (was) 128%.'

Word quickly went around the committees that a good way of testing the bona fides of claimants was to ask what they remembered of the Night of the Big Wind, which laid waste to Ireland and Britain in January 1839. Seventy years further on, in 1989, Lisa Shields and Denis Fitzgerald of Met Éireann revisited the events of 1909 and 1839. They unearthed the memory of a Galway storyteller, Tomás Laighleis, who was taken out of his schoolroom by the parish priest in 1909 and told to round up the elderly of the parish for a memory test.

The dreadful storm that rampaged across Ireland on the night of 6–7 January 1839 seared itself deep into the folk consciousness. A poem of twenty-five verses, '*Óiche Na Gaoithe Móire, ná Deireadh An tSaoil*', penned by the Galway poet Michael Burke, was still a favourite party piece a century later. Translating as 'The Night of the Big Wind, or The End of the World', the poem said that the 'night of storm and burning' would be 'clearly remembered for ever'.

The Dublin Evening Post agreed. Reporting on the devastation, the paper said:

> Comparing it with all similar visitations in these latitudes, of which there exists any record, we would say that, for the violence of the hurricane and deplorable effects which followed, as well as for its extensive sweep, embracing as it did the whole island in its destructive career, it remains not only without a parallel, but leaves far away in the distance all that ever occurred in Ireland before. With the exception of the frightful disasters [in Britain] the sister island appears to have escaped with comparative good fortune ... Ireland has been the chief victim of the hurricane – every part of Ireland, every field, every town, every village in Ireland have felt its dire effects.

The same newspaper described the capital as 'a sacked city', with roofs and chimneys ripped up and tossed about by the wind. Dundalk and Newry were hit particularly badly, with reports suggesting that nearly every roof in the latter had been torn away. In Limerick, the *Chronicle* reported that 'not a public edifice or institution in the city escaped the ravages of the storm and the best-built houses ... were sadly dismantled in the upper storeys.' *The Galway Patriot* said that seven people had been killed by collapsing masonry and chimneys, while the *Ballyshannon Herald* said that the population of Donegal were so terrified that many believed that the end of the world really was nigh.

The *Kerry Evening Post* reported that some of the villagers of Ventry were convinced that 'the priest's curse had brought it down' on them to punish the wickedness of those who had forsaken their faith. The faithful of the Bethesda Chapel on Dublin's Dorset Street may also have felt a little forsaken by their God. Hours before the storm struck they had given thanks at Sunday service for their safe delivery from an outbreak of fire the previous day. Sadly, the fire they believed extinguished may have been fanned back to life by the gale, and that night a fresh blaze destroyed the chapel, along with its female penitentiary and orphanage.

All across the land, thatched roofs went up in flames as embers from hearths were fanned and flung about by the gusts. The *Tuam Herald* carried reports of fire raining from the skies. In Loughrea, County Galway, 71 buildings were razed to the ground while in Athlone more than 100 burned down.

It wasn't just burning embers raining from the skies. The *Kerry Evening Post* told of 'a fine specimen of that rare and beautiful bird, the Stormy Petrel' which was cast to the ground in Westmeath, over ninety miles from its fishing grounds. Years later, *The Catholic Bulletin* published a recollection that 'all along the west coast for many days afterwards herrings were found six miles inland'. Two separate reports from Cavan told of fish being plucked up from lakes and dropped in far off fields. According to *The Dublin Evening Post*: 'Trees, ten or twelve miles from the sea, were covered with salt brine, and in the very centre of the Island, forty or fifty miles inland, such vegetable matter as it occurred to individuals to test had universally a saline taste.'

The storm raged so hard in parts that it physically rearranged the landscape. The *Ballyshannon Herald* reported: 'The sand banks at the bar to our harbour were so considerably lowered by the storm that ... the sea washed over the "sugar-loafs" and boats could pass where the tide had not before reached within half a mile.' The *Limerick Chronicle* told of how 'three acres of the bog at Glounamuckalough' were yanked up like a strip of carpet. The bog 'moved completely from its position and after traversing a distance of a mile, and crossing a rapid river, landed on the opposite side. Not an atom of surface is to be seen where the bog left but mere yellow mould. The occurrence fortunately has not done any injury to horses or cattle but is an incalculable loss to the owner of the land, Mr James Barry, as the bog rests on the very best portion of his farm.'

Other members of the landed gentry may not have suffered the misfortune of having three acres of bog dumped on their prime land, but many saw their wooded estates devastated. *The Dublin Evening Post* pined: 'The damage which it has done is almost beyond calculation. Several hundreds of thousands of trees have been levelled to the ground. More than half a century must elapse before Ireland, in this  regard, presents the appearance she did last summer.' The *Tuam Herald* reported: 'Trees of fifty and sixty years old were snapped in pieces like glass.' *The Dublin Evening Post* added: 'One of the extraordinary consequences which has followed this calamity [is] that timber is now a drug in Ireland, absolutely a drug. We know from an eminent timber merchant that a lot of trees which he would have gladly bought for £300 on Saturday, he purchased for £40 on Monday … No timber merchant in Dublin would go ten miles out of the city to buy a tree, and he would not, just now, take a present of 1,000 twenty miles from Dublin … There is a glut, and these beautiful trees are now nearly valueless.'

The suffering of the rich, as ever, paled against the suffering of the poor. *The Dublin Evening Post* empathised: 'The poor of course, as being the most numerous, have been the greatest sufferers. Tens of thousands of their wretched cabins have been swept away or unroofed – and many, as we have seen, have become a prey to the flames.' Sir William Wilde, father of Oscar, noted that with their winter store of fodder destroyed by flame or scattered on the winds, many poor people were faced with the desperate choice between slaughtering their cattle or keeping them alive with the potatoes stored to feed the family. Wilde wrote: 'The latter was resorted to by most of the poor, although they had a certainty before them that ere the new crop would come in, they themselves would be destitute.'

As the dust began to settle on the Night of the Great Wind, some commentators decided that the only proper response was to shoot the messenger. For over 100 years the members of the Royal Dublin Society had been advancing better and more scientific ways of measuring and forecasting the weather. Now *The Dublin Evening Post* took a swipe at the Society, saying that if it was 'worth a straw' it would furnish the Irish public with 'a complete and minute account of the mischief' and of any law of nature, if one exists, 'which guides the whirlwind and directs the storm'.

Stung by this criticism, the Society took out a newspaper advert inviting members of the general public to send details of storm damage to its officer at Dublin Castle. Interestingly, the advert stressed that the RDS would only welcome correspondence 'with real signatures'.

PILLS, THRILLS AND WEATHER BALLOONS

The World's First Air Disaster Strikes Tullamore

IN 1752, LONG BEFORE HE WAS ELEVATED to the role of American Revolutionary hero, the 46-year-old Benjamin Franklin conducted one of the most famous weather experiments in history. On a dark June afternoon in Philadelphia, Franklin embarked on a kite-flying exercise that he hoped would prove that lightning from the skies was an electrical phenomenon, and that he could channel that energy into a current of electricity – electrical currents being the latest hot science.

With the help of his son, Franklin attached his kite to a silk string, tying an iron key at the other end. Next, they attached a thin metal wire to the key and inserted the wire into a Leyden jar, which was a rudimentary form of battery that could store an electrical charge. As the sky blackened and a thunderstorm approached, Franklin sent the kite soaring on the wind and, once it was aloft, retreated into a barn so that he would not get wet.

The thunderclouds engulfed Franklin's kite which attracted the negative charges, which passed down the wet silk string to the key and finally into the jar. When he moved his free hand close to the iron key he received a small shock as the negative charges in the key were strongly attracted to the positive charges in his body, causing a spark to jump dramatically from the key to his hand. Franklin's experiment had proved that lightning was, in fact, static electricity.

He didn't appreciate it at the time, but Benjamin Franklin was lucky to have survived. News of his experiment spread rapidly, leading others to follow in his footsteps. Several were zapped to death. Two years earlier Franklin had speculated that this dangerous form of electricity could be channelled harmlessly into the ground by the use of a suitable conductor. To this end, he is widely credited with inventing the first lightning rod, although there are many prior claims from Sri Lanka to Russia. In the years after his kite experiment he began publicly advocating the use of lightning rods to protect private homes and public buildings. He found many takers. Years later, lightning struck Franklin's house but his clever invention kept the roof over his head intact.

Kites would make a comeback as a tool of weathermen towards the end of the nineteenth century, but for a long stretch in between they would be overtaken by balloons as the best option for scaling new scientific heights. During the Age of Enlightenment the new breed of curious individuals calling themselves scientists were making it up as they went along, with the result that the meticulous analyst and the madcap adventurer might equally turn up something for the books.

Some of the most madcap adventurers were the balloonists, many of whom combined a thirst for knowledge about the winds and the sky with the thrill-seeker's lust for fame and glory.

In June 1783 the Montgolfier brothers gave the first public demonstration of a manned balloon flight, rising 2,000 metres above the ground and travelling some 2km in ten minutes. The balloon's potential as a way of exploring the skies was made clear in an instant.

Eighteen months after the French siblings' maiden flight in a hot air balloon, Wicklow native Richard Crosbie made the first successful manned flight in Ireland, this time using hydrogen for lift. Hydrogen eliminated the need to keep a furnace stoked to generate hot air, which was a major fire hazard.

Richard Crosbie was both a gifted scientist and an incorrigible thrill-seeker. His ambition was to become the first person to fly across the Irish Sea, but ballooning was an expensive pursuit. To raise money he put what he called his Aeronautic Chariot on exhibition to Dubliners. The chariot was a basket specially modified to carry himself, his food, ballast, and his scientific equipment on an extended flight. He also flew a balloon over Ranelagh Gardens on the outskirts of Dublin, sending up a different animal every day. To test the winds that he hoped would carry him to Britain, he set loose a balloon carrying a cat which was spotted crossing the west coast of Scotland but then did a dramatic U-turn and splashed down into the sea close to a ship passing the Isle of Man which recovered both the aircraft and the animal.

In January 1785 large crowds turned out at Ranelagh Gardens to watch Crosbie make the first manned flight over Irish soil. Such was the interest that many arrived to discover that their tickets were fakes, while the approaches were gridlocked with the horse-drawn carriages of the well-heeled. *The Annual Register* glowed: 'The balloon and chariot were beautifully painted, and the arms of Ireland emblazoned on them in superior elegance of taste.' Crosbie was dressed for the high-altitude cold: 'His aerial dress consisted of a robe of oiled silk, lined with white fur, his waistcoat and breeches in one, of white satin quilted, and morocco boots, and a montero cap of leopard skin.'

In the end, Crosbie's landmark flight stopped short of even reaching the Irish Sea. He did make it across the River Liffey, though, landing to the cheers of bystanders on the North Strand at Clontarf.

Just four months later the County Offaly town of Tullamore was the scene of the world's first major air disaster when a balloon came down in flames on the tight cluster of streets, razing some hundred homes to the ground. Details of the event are sketchy, with contemporary newspaper reports saying it had been launched from the yard of a Doctor Blakely, probably on the outskirts of the town. Two years later John Wesley, the founder of Methodism and a regular visitor to Tullamore, asked in his diary: 'Have all the balloons in Europe done so much good as can counterbalance the harm which one of them did here a year or two ago? It took fire in its flight and dropped it down on one and another of the thatched houses so fast that it was not possible to quench it, til most of the town was burnt down.'

The great fire of Tullamore, which required the town to be substantially rebuilt, did nothing to quell the general enthusiasm for ballooning as a means of investigating the weather and showing off.

Almost three decades after Richard Crosbie's attempt to balloon across the Irish Sea fell far short of the mark, an English chemist named James Sadler made another attempt. In October 1812 Sadler took off from the grounds of Belvedere House in Dublin's Drumcondra. Two military bands and a large crowd turned up to see him cast off in a brightly coloured dirigible called *Erin go Brágh* (*Ireland Forever*). Sadler's balloon reached the airspace over the Welsh headland of Anglesey, where he could have landed and claimed the sea-crossing record. However, Liverpool was his original destination and he decided to press on towards the northwest. It was a big mistake. The winds carried him back out

A to Z
of Irish Weather

Kelvin The kelvin is a unit of measurement of temperature. It is named after the Belfast-born engineer and physicist William Thomson (1824–1907) who determined the correct value of Absolute Zero. He was created the first Baron Kelvin by Queen Victoria for his pioneering work on the first transatlantic telegraph cable which was laid from Valentia Island in County Kerry to Newfoundland, and for his staunch opposition to Irish Home Rule.

Krakatoa The 1883 eruption of Krakatoa in the Dutch East Indies (modern Indonesia) began in May and climaxed in August with a number of huge explosions that destroyed much of the island and its surrounding archipelago. The official death toll recorded by the Dutch was 36,417, but some authorities put this closer to 120,000. The average global temperature dropped by over 1 °C the following year and weather patterns were disrupted for several years more. The soot in the air darkened skies around the globe for years, producing spectacular sunsets. Ettie French, the daughter of songwriter, painter and inspector of County Cavan's drains Percy French, wrote that in the year after Krakatoa erupted her father was 'bowled over' by a series of 'wonderful sunsets' in the Irish midlands. Ettie noted: 'He went out every evening and tried to capture in paint the colours, which were due to volcanic dust.' Some art critics have argued that the startling crimson sky of Edvard Munch's 1883 painting *The Scream* was inspired by the disturbed atmospherics over Norway, while others beg to differ.

A TO Z
OF IRISH WEATHER

Lightning The proposition that lightning never strikes twice in the same place is plain wrong, as is amply demonstrated by the case of Roy Cleveland Sullivan (1912–1983), a ranger in Shenandoah National Park in Virginia, USA. Between 1942 and 1977, Sullivan was hit by lightning seven times, which earned him an unwanted place in *The Guinness Book of World Records*. For his troubles, Sullivan was nicknamed 'The Human Lightning Rod' and fair-weather friends tended to avoid his company whenever storm clouds gathered.

to sea where he was forced to ditch, denying him the official first crossing. Sadler's son Windham did make the first flight across the Irish Sea in 1817, although he was to die seven years later, still in his twenties, when he was thrown from a balloon.

Richard Crosbie of Baltinglass in Wicklow had introduced Ireland to the ballooning craze in the 1770s, and a couple of generations later another Baltinglass native, Benjamin O'Neale Stratford, attempted to take the science of ballooning onto a higher plane. Stratford, in fact, was the Baron of Baltinglass, and in 1834 was elected a life member of the prestigious Royal Dublin Society which was founded 'to promote and develop agriculture, arts, industry and science in Ireland'.

The scientist and the egotist in Stratford combined to set him the challenge of building the biggest balloon the world had ever seen. Around the time he joined the fellowship of the RDS, he had a giant hangar constructed on the grounds of his Wicklow estate, spending the next twenty years working on his project in total secrecy. He was so fearful of rivals that he kept only one faithful servant. The paranoid Stafford wouldn't even employ a cook, so he had his meals ferried in every day by Royal Mail coach.

By 1856 the giant balloon was almost complete. Stratford had even bought a landing strip on the banks of the Seine in Paris for his maiden flight. Then disaster struck. Stratford Lodge went up in flames. The locals arrived to tackle the blaze, but he reportedly diverted the firefighters away from his stately home, telling them to save the balloon at all costs. Their rescue efforts failed.

Stafford put in a compensation claim with the local authority, claiming the fire was an act of arson. A withdrawn loner for two decades, he had few friends amongst his neighbouring big farmers and businessmen, who warned him in court that the arson was imagined and he wouldn't see a penny from their community chest. A broken man, Stratford left for Alicante in Spain where he lived reclusively in hotels, rarely leaving his room. He would have his meals sent up by room service, but he would never allow the dirty dishes and cutlery to be collected. When his suite was full of dirty crockery he'd simply move to another. With his stately home and his fortune gone, he supplemented his income by breeding dogs and selling the hugely popular, but utterly useless, Holloway's Pills, which claimed to cure everything from blotchy skin, to asthma to tumours.

While the unfortunate Stratford was reduced to flogging his dodgy pills in Spain, others were pressing ahead with the scientific exploration of the atmosphere beyond the clouds and the effects it was having on the weather. For decades after the fire at Stratford Lodge the best way of charting the conditions at high altitude was for an aeronaut to ascend as high as possible in a balloon. The obvious shortcomings of this method included the fact that it was dangerous and that humans could not breathe, nor balloons fly, above a certain altitude.

In 1883 the cataclysmic eruption of the Krakatoa volcano in the Dutch East Indies sent millions of tons of soot into the upper atmosphere, darkening the skies around much of the globe. The distribution of this dark matter led weather watchers to speculate about the existence of fast-moving air currents high up in the atmosphere. Today we know these currents as the jet stream, but at the time of Krakatoa they were labelled 'the equatorial smoke stream'.

A few years later, in the 1890s, a new instrument called the robotic meteograph was invented which could measure weather conditions at the high altitudes of the jet stream without the need for human input. For a decade or so, weather kites made a storming comeback. Box-kites with meteographs attached could now go much higher than any manned balloon, and the science of weather took a leap forward. The kites were tethered to the ground, which put a limit on how high they could climb, and by the first years of the twentieth century unmanned hydrogen balloons were back in vogue, carrying the measuring boxes to new heights.

The primitive meteograph gave way in the 1920s to the more sophisticated radiosonde, meaning 'radioprobe'. Almost a century after its invention, the radiosonde remains a key component of modern weather analysis and forecasting. Today thousands of radiosondes attached to latex balloons are sent into the upper atmosphere every day from some 800 launch sites around the globe, with the Valentia Island weather station in County Kerry a key point in the network. The weather balloons calculate, cross-match and transmit data on air pressure, altitude, latitude and longitude, temperature, relative humidity, wind speed and direction, and the strength of incoming cosmic rays.

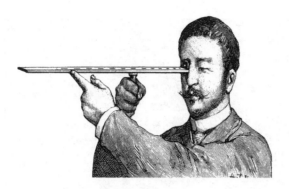

THE MEASURE
OF A MAN

Sir Francis Beaufort (1774–1857)

O N A CRISP MORNING IN JANUARY 1806, the British Navy
frigate HMS *Woolwich* sailed from the Isle of Wight on
a voyage to conduct a hydrographic survey of the Rio
de la Plata estuary in Central America. Taking charge of his first
ship's command was 32-year-old Francis Beaufort from Navan in
County Meath. That evening, as his ship passed south of Ireland,
Beaufort wrote in his log: 'Hereafter I shall estimate the force of the
wind according to the following scale, since nothing can convey
a more uncertain idea of the wind than the old expressions of
moderate and stiff etc.'

As one of the Royal Navy's highly trained hydrographers at
a time when Britain ruled the waves unrivalled, Beaufort's job
was to make navigation safer by mapping the topography of the
open seas, inlets, bays and great rivers. This involved measuring the
depths, currents and tides, while charting reefs, landmarks and other

A TO Z
OF IRISH WEATHER

March The third month of both the Julian and Gregorian calendars, March was the first month of the earliest Roman calendar. Originally called Martius, it was named after Mars, the god of war and agriculture. March for the Ancient Romans was the month when both farming and warfare could restart after the winter.

May The fifth month of the Julian and Gregorian calendars is named for Maia, the Greek goddess of fertility, nurturing and growth.

Met Éireann From 1860, when the first weather data was telegraphed from Valentia Island in County Kerry to the British Met Office in London, Ireland's weather stations were integrated into a forecasting network covering the British Isles. Although the Free State became independent in 1922, Britain continued for years after to administer Ireland's weather service. When commercial transatlantic flying became a practical reality in the mid-1930s, Foynes Airport on the Shannon Estuary took on a huge strategic importance as a key stepping stone for flights refuelling between Europe and North America. In 1936 the Irish, British and Canadian governments entered into an agreement sharing a form of joint authority over the landing rights at Foynes, which would be replaced by Shannon Airport in the 1940s.

Part of the new deal was that the Free State would outfit Foynes with the most up-to-date wireless and meteorological equipment that money could buy. In July 1936 the acting Minister for Industry & Commerce, Thomas Derrig, told the Dáil: 'It is, I think, generally known that the rather limited meteorological service at present in existence in the Saorstát [Free State] has hitherto been maintained by the United Kingdom Government. Arrangements are being made for the transfer of this service to the control of the Saorstát Government, and it is our intention to develop it into a first-class meteorological service, and I may say incidentally at this point that the United Kingdom Government has agreed to contribute £6,000 per annum towards the cost on the basis of services rendered.'

The following year the new Irish Meteorological Service took over the running of Ireland's weather stations from the British Met Office, which continued for some years to subsidise the Irish service and to send over personnel to provide necessary expertise.

Microclimates A microclimate is a localised zone where the climate differs from the surrounding area. Ireland has several distinct microclimates. Warmed by the Gulf Stream, parts of coastal west Cork support subtropical vegetation, with Garnish Island a prime example. Killary Harbour in County Mayo and Newcastle in County Down have notably drier microclimates than their hinterlands. Perhaps the best-known microclimate belongs to the Burren area of County Clare, which is kept unusually temperate by the Gulf Stream, allowing it to support a range of plants and animals not tolerated elsewhere by the Irish climate.

Merrow The Merrow (*murbhach* in Irish) was a mermaid believed by fishermen to appear in advance of a coming storm. In his 1888 compendium, *Fairy And Folk Tales of the Irish Peasantry*, W. B. Yeats wrote: 'It was rather annoying to Jack that, though living in a place where the merrows were as plenty as lobsters, he never could get a right view of one.'

Mood Molecules of air contain electrons which, depending on the conditions, can carry a positive or negative charge. These positive and negative ions which we inhale in their countless billions, are associated with different places and weather conditions, and are credited with producing different moods in humans. Negative ions are most abundant in the bracing air high on mountains, on sea coasts, and close to waterfalls. The smashing of water droplets into the spray of a waterfall produces feel-good negative ions, and by the same token the spray produced by a bathroom shower is credited with inducing a more refreshing, wide-awake effect than a still-water bath. The reason cited for this feel-good factor is that once the negative ions reach our bloodstream, they produce biochemical reactions that increase levels of the mood chemical serotonin, helping to fight depression and stress and to boost energy levels. On the other side of the coin, positive ions associated with thunderstorms are believed to produce agitation and anxiety in susceptible people.

features. Mapping was in the Beaufort blood. The commander's father, Dr Daniel Beaufort, took time out from his duties as Navan's church rector to chart and publish a new and much improved map of Ireland in 1792.

Severely wounded during a navy engagement off Malaga in 1800, Beaufort returned home to recuperate, but found himself drawn into yet more measuring by his brother-in-law Richard Lovell Edgeworth. Richard Edgeworth, father of the celebrated novelist Maria, was an inventor and man ahead of his time whose creations included a machine to measure the size of a plot of land, and a prototype caterpillar track now found on bulldozers and tanks, which he described in the 1770s as 'a cart that carries its own road'. With the invasion of Ireland by Napoleon a real possibility, Beaufort and Edgeworth spent two years constructing a semaphore line to transmit messages quickly between Dublin and Galway.

By the time he set sail from England in January 1806, Beaufort had become preoccupied with measuring the winds at sea. Since time immemorial, sailing ships had been at the mercy of the winds, but with no instruments available to measure wind strength accurately, the entire sphere was a muddle of guesswork and tall tales. As he noted on Day One of his voyage log, the Irishman was determined to come up with a more scientific gauge of the winds which would measure force rather than speed. He realised that it was impossible to measure the speed of the wind from the deck of a ship in motion, but that a quick look at the way the sails were billowing, or not, could be made to fit a visual scale.

The Beaufort Wind Force Scale, for which he would become famous, borrowed heavily on the work of the Royal Navy's first hydrographer Alexander Dalrymple, who in turn had honed an idea by the windmill expert John Smeaton. This easy give-and-take for the greater good of the newly forged entity of Great Britain might explain why Dalrymple championed Beaufort in an 1808 letter to the Admiralty praising his survey of the Rio de la Plata with the commendation: 'We have few officers (indeed I do not know one) in our Service who have half his professional knowledge and ability, and in zeal and perseverance he cannot be excelled.'

Beaufort devised his wind scale to describe the effects of the wind on a fully rigged man-of-war frigate in the service of the Royal Navy. A key element in his calculations was the amount of sail a ship could raise on its masts in high winds without causing itself damage. The first point on his scale, 0, is entitled 'calm' and describes the state of the sea as 'like a mirror'. Next on the scale is 1 meaning 'light air', which translates as 'ripples, no foam'. Midway, at 6, we have 'strong breeze' which is described as 'some large waves, extensive white foam crests, some spray'. The most feared number is 12 or 'hurricane', which conjures up a nightmare of 'air filled with foam and spray, sea white, visibility extremely bad'.

Just as he had modified earlier prototypes for his scale, there have been further tweaks over the past two centuries. With the end of the sail era and the coming of steam propulsion, Beaufort's scale was shifted from measuring the effects of winds on ships to their impact on the waves themselves. The man who adapted the Beaufort Scale for the steam era at the beginning of the twentieth century, Sir George Simpson, was approached in the 1920s by the International Meteorological Committee (IMC) and asked if he'd do his trick again. With the global village fast becoming a reality, the IMC asked Simpson to simplify the Beaufort Scale for landlubbers around the world. Simpson replaced Beaufort's marine symbols with new ones, including rising smoke, swishing trees and rustling leaves, and he introduced measured wind speeds.

Having given his name to his new and improved weather scale while still a young man, Francis Beaufort devoted the remainder of his long and distinguished life to ensuring it was taken aboard Royal Navy culture. He hit the standard retirement age of fifty-five in

1829, but far from dropping anchor on a glorious career he found a second wind. Appointed to the grand post of British Admiralty Hydrographer of the Navy, he set about using his clout to launch many pioneering scientific seaborne endeavours.

One of Beaufort's protégés was Robert FitzRoy, who was given temporary command of the survey vessel HMS *Beagle* after her captain committed suicide. Together with the Earl of Grafton (of the family that gave its name to Dublin's flagship shopping street), Beaufort secured FitzRoy as the new full-time captain of the *Beagle*. With a fresh voyage of exploration in the offing, FitzRoy asked Beaufort if he knew of any 'well-educated and scientific gentleman' who might go along for the ride. Beaufort approached several, but there were no takers until a young botanist called Charles Darwin answered the call.

The rest, as we know, is history.

THE OTHER
POTATO FAMINE

The Great Frost and
The Year of Slaughter 1740–41

WHEN IRELAND FROZE HARD in the first days of 1740 following a storm of rare ferocity, the novelty of the so-called Great Frost was an excuse for celebration across the land. *The Dublin Evening Post* reported a large open-air ball held on the surface of the River Boyne which was frozen solid, where local grandees participated in 'several country dances on the ice, being attended by a large band of music'. Reports told of whole sheep being roasted for revels on the River Shannon which was thick with ice to a depth of nineteen inches, and of traders strolling to market across the frozen face of Lough Neagh in Ulster.

The party mood didn't last as the cold snap soon outstayed its welcome and began to bite hard. In his superb account of the crisis, *Arctic Ireland*, historian David Dickson suggests that Ireland's urban dwellers were the first to get a taste of the hardships to come. While rural cottage dwellers would have laid up a stock of turf for the winter, those living in Dublin and the other towns along the eastern seaboard had become reliant on British supplies of coal, which burned hotter and was considered more upmarket. With the ports frozen for most of January, the coal supply dried up. When the ice eased at the end of the month and imports resumed, the price of coal spiked to the point that it became unaffordable to many. This in turn sparked an orgy of illegal tree chopping, with fourteen individuals arrested for felling trees in the Phoenix Park alone. The coal merchants were widely blamed for stockpiling and profiteering.

The extreme cold snap was not the first to be dubbed the Great Frost during the prolonged cool period known as the Little Ice Age, which affected Europe from about 1600 to 1850. Other winters given the same nickname in the British Isles and Europe included 1607–8, 1683–4 and 1708–9. Although scientists had begun serious attempts to understand the cause and effect of weather conditions, by 1740 there was no rational explanation for the great storm immediately followed by an Arctic deep freeze that brought extreme cold but no snow. Some, much later, would speculate that the world's climate was thrown into a delayed-action chaos by two megathrust earthquakes on the volcanic Russian peninsula of Kamchatka in October 1737. Megathrust earthquakes occur at the Earth's subduction zones where one tectonic plate is forced beneath another, and are the most violent forces on the planet's surface. However, as 1740 wore on, the people of Ireland were less concerned with the cause of the misbehaving weather than with its deadly effects.

A series of failed grain harvests in the 1720s had pushed the poor into an ever deeper dependence on potatoes, which by 1740 had become the dietary mainstay of much of the population for most of the year. Before the heavy frost had even lifted, a

Cork estate agent called Richard Purcell noted that the potatoes were frozen in the ground and in storage, and he penned a chilling prediction: 'The eating potatoes are all destroyed, which many think will be followed by a famine among the poor, and if the small ones, which are not bigger than small peas and which be deepest in the ground, are so destroyed as not to serve for seed, I think this frost the most dreadful calamity that ever befell this poor kingdom.'

By the end of February the frost had finally vanished, only for the weather to confound and dismay with another strange twist. The big freeze continued, but it was a bone-dry freeze. The rains that normally drench Ireland throughout the year shied away and the countryside became chapped with an Arctic drought.

As spring turned to summer the temperature struggled upwards, but there was little respite from the drought, which was felt over a large part of Europe. The failure of the potato crop caused a big jump in demand for grain, but the drought severely damaged the grain harvest so that the prices demanded for wheat, oats and barley skyrocketed. For the ordinary person, the soaring grain prices could be directly measured in the shrinking size of the loaf of daily bread. As they had often done before in times of desperation, the poor took to eating nettles as a proven source of nourishment. In Dublin, Drogheda and other towns, there were food riots and the looting of shops and storage depots.

The weather played more cruel tricks as the Christmas of 1740 neared. A series of blizzards in October and November were followed by a torrential deluge of rain in early December which in turn was followed by a sudden deep freeze. The millions of tons of water locked up by the big freeze were then liberated by a sudden sharp temperature rise which caused flash floods, propelling deadly miniature icebergs into the paths of boats and buildings.

The catastrophe came to a horrible head in the spring and summer of 1741, which became known to the people as *Bliain an Áir*, or the Year of Slaughter. Starvation compounded a range of diseases and led to epidemics of 'flux' (dysentery), typhus and other ills.

In April of that year, Sir Richard Cox, son of the former Lord Chancellor of Ireland, wrote despairingly: 'Mortality is now no longer heeded, the instances are so frequent. And burying the dead, which used to be one of the most religious acts among the Irish, is now become a burthen: so that I am daily forced to make those who remain carry dead bodies to the churchyards, which would otherwise rot in the open air. Otherwise I assure you the common practice is to let the tree lie where it falls, and if some good natured body covers it with the next ditch, it is the most to be expected. In short, by all I can learn, the dreadfullest civil war or most raging plague never destroyed so many as this season. The distempers and famine increase so that it is no vain fear that there will not be hands to save the harvest.'

The weather through to the autumn of 1741 remained out of sorts, and the grain and potato harvests were again poor, but there was a noticeable improvement. The real turning point came in July when tons of old wheat hoarded by merchants were released onto the market. The loaf of bread once again became affordable, and the healing process could begin.

Historian David Dickson, the leading authority on the crisis of 1740/41, estimates that up to 20 per cent of the population, or 480,000 people, may have perished between the Great Frost and the Year of Slaughter.

IRISH WEATHER PROVERBS
— BIRDS

A storm will come when seagulls fly inland.

If the robin stands at the doorstep, the weather will be foul.

When the raven flies high in the sky the days will be wet and windy.

A storm will blow if ducks stand up and ruffle their feathers.

When wild geese honk at night the weather will be delightful.

When the wild geese fly south to the coastal cliffs, snow will fall.

The shrill oystercatchers bring wet and windy days.

Tomorrow will be a miserable day if the hens stay out after dark.

When the heron swims upstream by the mountains, the weather will be dry but rough. When she goes downstream it will rain.

If the cock crows on a rainy day, the weather will clear up.

The curlew cries when rainclouds fill the sky.

When the robin sits on the highest branch the days will be fine.

The cry of the snipe will bring frost at night.

The three coldest days of the year are the days that killed the stonechat.

Low flying swallows bring bad weather.

If the blackbirds call at dusk there will be a sharp chill in the night.

When the hen picks her plumage a downpour is coming.

The great northern diver calling into the haven means frost.

WHEN WEATHER
CHANGED
HISTORY
PART 6

Wolfe Tone's French Armada of 1796

SEVEN YEARS AFTER THE FRENCH REVOLUTION of 1789, the government of the French First Republic assembled an invasion force of some thirty-three ships at the port of Brest. Their destination was Bantry Bay in County Cork where they hoped to land 15,000 troops and light the touchpaper on a general insurrection against British rule in Ireland. If the revolt proved successful, the French hoped that Ireland could be used as a stepping stone for an invasion of Britain.

On board the *Indomptable* was Theobald Wolfe Tone, the revolutionary figurehead of the United Irishmen, a body dedicated to the setting up of an Irish Republic. Launching an invasion by sea in December was tempting fate, and fate responded most cruelly. The weather that winter would turn out to be the worst on record since 1708, almost a century earlier. From the time the fleet left France it was lashed by storms, with many ships scattered and some sunk.

Wolfe Tone kept a personal log of his voyage, and it provides an eloquent record of a doomed venture. On 15 December 1796, having weathered the first storms, he wrote: 'It is the most delicious weather, and the sun is as bright and warm as in the month of May … We are all in high spirits and the troops are as gay as if they were going to a ball.' Two days later he was still in buoyant mood, writing: 'This day has passed without any event; the weather moderate, the wind favourable and our eighteen sail pretty well together.'

It wasn't to last. By the next day, 18 December, Tone was tormented with the possibility that the French fleet wouldn't turn up intact, while the British fleet might appear in hot pursuit any minute. He wrote: 'At nine this morning a fog so thick that we cannot see a ship's length before us. "Hazy weather, master Noah." Damn it, we may be, for aught I to know, within a quarter of a mile of our missing ships, without knowing it. It is true we may also, by the same means, miss the English, so it may be good as well as evil, and I count firmly upon the fortune of the republic.'

That night he added: 'This damned fog continues without interruption. Foggy all day, and no appearance of our comrades. I asked General Cherin what we should do, in case they did not rejoin us. He said that he supposed General Grouchy would take the command with the troops we had with us, which, on examination, we found to amount to about 6,500 men. I need not say that I supported this idea with all my might.'

The dead calm that fell on the invasion force was as disruptive as the heavy storms. On 19 December, Tone wrote in utter frustration: 'I see about a dozen sail, but whether they are our friends or enemies

God knows. It is stark calm, so that we do not move an inch even with our studding sails; but here we lie rolling like so many logs on the water. It is most inconceivably provoking; two frigates that were ordered to reconnoitre have not advanced one hundred yards in an hour, with all their canvas out. It is now nine o'clock; damn it to hell for a calm, and in the middle of December … This calm! This calm! It is most terribly vexatious.'

On 21 December, after enduring more squalls, the weather suddenly changed for the better. The invasion could now proceed. The only thing missing was most of the invasion force. A hint of resignation was creeping into Wolfe Tone's thoughts as he wrote: 'At day break we are under Cape Clear, distant about four leagues, so I have, at all events, once more seen my country; but the pleasure I should otherwise feel at this is totally destroyed by the absence of the General who has not joined us and of whom we know nothing … It is most delicious weather, with a favourable wind, and every thing, in short, that we can desire, except our absent comrades.' Worse, the French officers on board didn't seem too pushed either way. Tone noted: 'When they speak of the expedition, it is in a style of despondency; and when they were not speaking of it they are playing cards and laughing … I am now so near the shore that I can see distinctly two old castles, yet I am utterly uncertain whether I shall ever set foot on it.'

That uncertainty grew when he penned his diary of 23 December, writing: 'Last night it blew a heavy gale from the eastward with snow, so that the mountains are covered this morning, which will render our bivouacs extremely amusing.' The invasion force, he noted despairingly, had now been flung asunder 'for the fourth time'. Of the remainder, he added: 'We are here, sixteen sail, great and small, scattered up and down in a noble bay, and so dispersed that there are not two together in one spot, save one, and there they are now so close that if it blows tonight as it did last night they will inevitably run foul of each other, unless one of them prefers driving onshore.'

A to Z
of Irish Weather

Newgrange Once a year, at the winter solstice, clouds permitting, the rising sun shines directly along the corridor of this 5,200-year-old passage tomb in County Meath, filling the main chamber with light for approximately seventeen minutes and, according to one commentator, 'making the stones glow like living gold'. Many experts argue that this is clear evidence of sun worship, and some speculate that the passage structure may have been designed to 'capture' the sun on the shortest day of the year, for reasons best known to its builders and their culture. For every theory there is a counter-theory and it's highly unlikely we'll ever know for sure.

Nuclear Winter. Also known as Atomic Winter, this theory posits that a nuclear war would ignite massive firestorms which in turn would blacken the skies with dense smoke and soot. The sun would be blocked out and bleak winter conditions would envelop the planet. During the particularly foul Irish summer of 1946 there was a great deal of speculation in the newspapers and around the dinner table that US nuclear tests on Bikini Atoll in the Pacific Ocean had brought on the bleak season by mutating the atmosphere.

On 24 December it seemed that Christmas had come a day early, as clearing weather allowed Tone and other officers to go on board the *Immortalité* and persuade General Grouchy to press ahead with the landings of the 6,500 men available. 'We are all as gay as larks,' he chirped.

Alas, Christmas Day brought bitter disappointment. Tone wrote despondently: 'Last night I had the strongest expectations that today we should debark, but at two this morning I was awakened by the wind. I rose immediately, and wrapping myself in my great coat walked for an hour in the gallery, devoured by the most gloomy reflections. The wind continues right ahead so that it is absolutely impossible to work up to the landing place, and God knows when it will change. The same wind is exactly favourable to bring the English upon us.

Tone's big fear now was that, even if the French did succeed in landing their troops, the English forces at Bantry, Cork and Limerick would by now be reinforced, and together with the approaching English fleet would trap them in a pincer movement.

He pondered: 'If we are taken, my fate will not be a mild one. The best I can expect is to be shot as an émigré rentre, unless I have the good fortune to be killed in the action ... This day at twelve the wind blows a gale, still from the east; and our situation is

A TO Z
OF IRISH WEATHER

October The tenth month of the Julian and Gregorian calendars, October was the eighth month of the old Roman calendar and retained its name Octo, meaning eight in Latin.

Ozone Layer/Ozone Hole In the spring of 1997 the renowned meteorologist Brendan McWilliams urged his *Irish Times* readers to go out and watch the skies for the ozone hole which would be passing over Ireland later that day. Some twigged that the date was 1 April – April Fool's Day – while others didn't. One unhappy skywatcher fumed on the letters page: 'I am livid with McWilliams.' The ozone layer is a band in the atmosphere where the scarce molecule ozone is found at its most highly concentrated. The ozone layer absorbs 97–99 per cent of the sun's medium-frequency ultraviolet light, shielding life forms from these potentially damaging rays. Chlorofluorocarbons (CFCs) from aerosols, refrigerators and other mod cons deplete the ozone in the atmosphere, and in 1985 a large ozone hole was discovered through which harmful ultraviolet-B radiation could penetrate to the Earth's surface. Scientists reported in 2003 that the phasing out of CFC production appears to have slowed the depletion of the ozone layer.

now as critical as possible, for it is mortally certain that this day or tomorrow on the morning the English fleet will be in the harbour's mouth, and then adieu to everything.'

The French commanders shared his fears and upped anchor for a hasty retreat back to France on 26 December. Down but not out, Tone consoled himself that he'd lived to fight another day, and put the disaster down to the capricious Irish weather. He reasoned: 'Notwithstanding all our blunders, it is the dreadful stormy weather and easterly winds, which have been blowing furiously and without intermission since we made Bantry Bay, that have ruined us. Well, England has not had such an escape since the Spanish Armada; and that expedition, like ours, was defeated by the weather. The elements fight against us and courage is of no avail. Well, let me think no more about it; it is lost, let it go!'

Wolfe Tone did fight another day, leading the United Irishmen rebellion of 1798. As the French attempted another seaborne invasion, again with no success, Tone was captured on board the disabled flagship *Hoche*. He committed suicide while awaiting trial. The failed expeditions of 1796 and 1798 were the last ever attempts by a Continental power to use Ireland for a back-door invasion of Britain.

A MIGHTY RIVER
IN THE OCEAN

The Gulf Stream, Sargasso Sea
and Ireland's Eels

THE SEA AROUND IRELAND is usually a good deal warmer around the southwest coasts of Cork and Kerry than it is off the northeast coast of County Antrim. In the middle of winter the average difference is a marked 3 °C, with the northern waters dropping to 7 °C compared to 10 °C in the south. These sharp variations have little to do with the fact that the top of the country is some 300 miles (480km) nearer the North Pole than the bottom, and a great deal to do with the warm Gulf Stream waters bathing the southwest.

The Gulf Stream is a mighty river in the ocean that arises in the Gulf of Mexico and makes its way at pace across the Atlantic. It carries a greater volume of water than all the rivers of the world combined. It is propelled by the rotation of the Earth (the Coriolis effect), by winds, and by the sun's warming of equatorial waters. As the water warms, its level rises, and it flows in a downhill fashion towards the lower levels of the surrounding ocean. This movement adds to the volume and force of the Gulf Stream. Just after midway through its journey it splits in two, with the Canary Current diverting south to bounce homewards off the coast of Africa, and the North Atlantic Drift heading northwards where it has a welcoming warming effect on Ireland and Britain in particular.

The first European to make note of the Gulf Stream was the Spanish explorer Juan Ponce de León who made landfall on a new territory while searching for the Fountain of Youth. He hadn't found the Fountain, but he did claim the new land for the Spanish king, naming it Florida. Days later de León ran into a current so strong that it forced his ships backwards and whisked one out of sight for two days. He noted the current in his log and Spanish captains quickly began hitching a lift as it was by far the handiest route home to Europe.

In geological terms, the Gulf Stream has only been with us a wet day. It was not until around 3 million years ago that the narrow strip of sea separating North and South America was closed by continental drift. When the Pacific Plate and the Caribbean Plate of the Earth's crust nudged up together, they profoundly changed the weather systems of the globe. Unable to flow into the Pacific any more, warm waters entering the Caribbean were diverted northwards into what we call the Gulf Stream.

More than 200 years after de León identified the northbound current, it fell to the great American polymath Benjamin Franklin to name it, although his preferred spelling was 'The Gulph Stream'. As Deputy Postmaster of Britain's American colonies, and the veteran of several Atlantic crossings, Franklin noticed that the journey time of ships from Ireland and Britain to America could vary enormously. Some routes could take up to a fortnight longer than others, although the distance wasn't too much different.

Franklin wrote to his cousin, Timothy Folger, for advice. A salty whaling captain, Folger knew the answer straight away. The

ships on the fast route were nipping across the Gulf Stream as soon as they came to it, like you might across a busy road, getting out of its Europe-bound current at the first opportunity and continuing on towards the New World. The slow ships were foolishly going against the flow, either through ignorance or sheer pig-headedness.

Franklin didn't rule out the latter. In 1785 he reported that his American whaling cousin complained of the snobbish response from the slow-going British captains: 'We ... advised them to cross it [the Gulf Stream] and get out of it, but they were too wise to be counseled by simple American fishermen.'

Franklin continued: 'When the winds are but light, [Folger] added, they are carried back by the current more than they are forwarded by the wind: and if the wind be good, the subtraction of seventy miles a day from their course is of some importance. I then observed that it was a pity no notice was taken of this current upon the charts, and requested him to mark it out for me which he readily complied with, adding directions for avoiding it in sailing from Europe to North America. I procured it to be engraved by order from the general post office, on the old chart of the Atlantic, at Mount & Page's, Tower-hill, and copies were sent down to Falmouth for the captains of the packets, who slighted it however; but it is since printed in France, of which edition I hereto annex a copy.'

Franklin went to a deal of trouble to have Folger's chart printed up for Britain's sea captains in 1769–70, and he was most miffed at the way it was rudely brushed aside. In 1776 he took on a key leadership role as the American colonies ignited into open conflict with Britain. With the Atlantic Ocean now a theatre of war, Franklin ordered the distribution of his Gulf Stream chart stopped in order to deny the enemy any edge.

Between the publication of the chart and its withdrawal, Ben Franklin paid a visit to Ireland where he was dismayed at the state of a land he took to be England's nearest and oldest colony. He was also struck by the mutinous atmosphere in the country. He wrote: 'In Ireland I had a good deal of conversation with the patriots; they were all on the American side of the question in which I am endeavor'd to confirm them. The lower people in that unhappy country are in a most wretched situation, thro' the restraints on their trade and manufactures. Their houses are dirty hovels of mud and straw, their clothing rags, and their food little beside potatoes. Perhaps three fourths of the Inhabitants are in this situation.'

He returned to the troubled colonies in 1775 and the next year was sent as envoy to Paris to enlist the support of the French for the American independence struggle. He made the most of the two transatlantic crossings to test the temperature of the Gulf Stream, confirming that it was warmer than the surrounding waters. From this he speculated that the Gulf Stream might have a warming effect on Europe. Established in Paris as the Ambassador of the thirteen colonies to the King of France, he succeeded in having his chart republished for the friendly French navy. During his long term as envoy to the court of Versailles, Franklin also came up with a daylight savings scheme, which, he claimed, would save Parisians a fortune in candle wax. (See p.186)

If the Gulf Stream can be described as a fast-flowing river in the ocean, the same can be said of the North Atlantic Drift and the Canary Current after they split off north and south towards Europe and Africa. Between them, the Stream, the Drift and the Current form the western, northern and eastern boundaries of the slowly rotating Sargasso Sea, which can be described as a pond in the ocean.

Lying a short way east of the Bahamas, the Sargasso Sea was first charted in the fifteenth century by Portuguese sailors who named it for the abundant Sargassum seaweed that clumped there in huge masses. In 1922 the Danish scientist Johannes

Ten Weather-Related Disasters At Sea

1645 *Great Lewis* During the English Civil War, Parliament sent the *Great Lewis* to re-provision the garrison in Duncannon Fort, County Wexford, which was under siege by Irish Confederate royalists. A treacherous tide and wind carried the ship into the firing range of the enemy. With its masts destroyed it drifted and sank, with the loss of 300 lives.

1656 *Two Brothers* In 1655 Oliver Cromwell's English forces had captured Jamaica from the Spanish. Cromwell's plan was to transport the most troublesome of his defeated Irish captives to Jamaica, offering them their freedom and 30 acres of land to colonise the large tropical island. The *Two Brothers* was transporting troops from Kinsale to Port Royal when it sank in Timoleague Bay, County Cork, with the loss of 241 lives.

1796 *L'Impatiente* The 44-gun frigate was part of a large fleet sent by the revolutionary French government to invade Ireland and bolster the revolt of Wolfe Tone's United Irishmen. It was smashed on the rocks at Crookhaven, County Cork, on the southwest coast of Ireland with the loss of some 540 men. (See pp.131–135)

1807 *Rochdale* and *Prince of Wales* Between them, the two ships were packed with troops bound for the war against Napoleon on the Continent. Large vessels entering or leaving Dublin Bay could do so only at high tide because of treacherous sandbanks, and the long wait to get out of harbour exposed them to sudden storms. Gale-force winds and snow blizzards sank the two ships with the combined loss of 400 lives. The tragedy prompted the building of Dun Laoghaire (Kingstown) Harbour as a safer haven.

1816 *Seahorse*, *Lord Melville* and *Boadicea* In January 1816 the three ships ran into a storm as they neared the end of their journey from Ramsgate in England to Cork. All were carrying troops returning from the Battle of Waterloo six months earlier. Battered by gales, the ships were dashed on the rocks of Tramore Bay, County Waterford, and Kinsale, County Cork. Around 340 people died on the *Seahorse* alone. Not one of the 33 women and 38 children aboard were amongst the 30 survivors. The storm took around 800 lives.

1853 *Queen Victoria* The paddle steamer was making its usual mail run between Liverpool and Dublin, carrying amongst its passengers a number of cattle dealers who had sold their stock in England. The ship ran into 'an impenetrable cloud' of snow off Dublin Bay, blocking out the light of the Baily Lighthouse on Howth Head. (Rumours circulated that the light was unlit.) Eight survivors somehow clung to the rocks of Howth Head, but eighty-three died.

1854 *Tayleur* Described as 'the first *Titanic*', the White Star Line clipper ran aground on Lambay Island in Dublin Bay on its maiden voyage from Liverpool to Melbourne, Australia. The ship's iron hull disabled the compasses, and the off-course vessel mistakenly sailed eastwards into a storm. Of over 650 on board only 290 survived.

1859 *Pomoma* The *Pomoma* left Liverpool carrying 373 passengers to New York. The ship ran into squalls in the Irish Sea, but there were suggestions from survivors that the reason it ran so far off course towards the County Wexford coastline was that the captain and crew were drunk. Passengers were confined below decks while crew escaped on lifeboats. The *Wexford Coroner* reported: 'We must express surprise that with a favourable wind and tolerable weather, the ship could have gone so much out of its course. We have no proof of drunkenness but most heartily we condemn that portion of the crew which deserted their passengers occupying the boats to the exclusion of women and children.'

1904 *Norge* The Danish passenger liner had picked up 800 passengers in Copenhagen, Oslo and Kristiansand for a journey to New York, when it ran aground on St Helen's Reef close to Rockall in the north Atlantic. The death toll was 635, while 160 survivors spent up to eight days in lifeboats. The captain, who had run his ship aground in clear summer conditions, was cleared of criminal charges.

1915 *Viknor* The British merchant cruiser was pressed into service at the start of the First World War. It disappeared off Tory Island on the Donegal coast in 1915 without sending a distress signal. The Germans had recently mined the area, but the cause may have been a violent storm, or a combination of the two. Of the 295 on board, none survived.

Schmidt first suggested that the Sargasso Sea could be the spawning ground for Ireland's eel population. Schmidt was attempting to solve a puzzle that had confounded Europeans for centuries. It was well known that millions of eels migrated down the rivers of the continent each year and out to the open sea, but no one knew for sure where they were headed. Schmidt's assertion that the Sargasso, 5,000 miles away, was their final destination met with some sniggers of derision, but time has vindicated him.

We now know that the eel spends its early years in the rivers of Ireland and Europe preparing for its marathon journey to the seaweed jungle in the mid-Atlantic. Once there, it is assumed it spawns and lays eggs; however, this has never been witnessed.

The eggs hatch into transparent larvae called leptocephalus and make the return journey floating along on the North Atlantic Drift. They begin the long trek as larvae, but by the time they reach Irish shores they have developed into tiny glass eels strong enough to swim up our rivers and into our lakes. Some eels grow to a metre in length before heading off to complete their circle of life.

In 2006, as part of an EU project to combat dwindling eel stocks, a team of investigators released twenty-two large adult eels into the sea off Galway. The eels were tagged with tiny pop-up satellite tags which, as the name suggests, detached from the eels after six months and popped up to float on the waves, sending back information on the journey to date. The data showed that rather than swim westwards towards the Sargasso Sea, the eels headed south past the Azores in order to pick up a conveyor belt ride on the North Equatorial Current heading westwards from Africa.

Starring Dennis Quaid and Jack Gyllenhaal, the 2004 disaster movie *The Day After Tomorrow* posits a scenario in which the thermohaline circulation of the Gulf Stream is disrupted by global

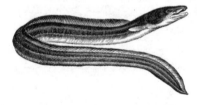

warming, contributing to an overnight glaciation of the Northern Hemisphere. While the Intergovernmental Panel On Climate Change (IPCC) has dismissed such a scenario as pure Hollywood fantasy, the UN-sponsored body has warned that some slowdown of the current is 'very likely' over the next century if the current level of greenhouse gas emissions continues unabated.

WHEN WEATHER CHANGED HISTORY PART 7

The Battle of Waterloo, 1815

I N HIS SWEEPING NOVEL *Les Misérables*, Victor Hugo asserted that the momentous Battle of Waterloo turned on the mood of the Belgian weather. Hugo wrote: 'If it hadn't rained during the night of June 17–18, 1815 the future of Europe would have been different. A few drops of water, more or less, brought Napoleon to his knees … a cloud crossing the sky out of season was enough for a whole world to disintegrate.'

Two centuries on from that crunch match between Napoleon and the great powers of Europe, experts still argue furiously as to whether the foul weather tipped the odds against the returning French Emperor, and in favour of the allied forces led by Dubliner Arthur Wellesley, the first Duke of Wellington. One of Wellington's most celebrated sayings is that Waterloo was 'a damn close-run thing', although that's not what he said. His actual words were 'a damn nice thing', using 'nice' in its old sense of 'uncertain', but either phrase underlines that the battle could have gone either way, and with it the shape of the world as we know it.

Victory at Waterloo made Wellesley the toast of Europe and catapulted him to the pinnacle of the power pyramid, including two spells as British Prime Minister (the second very brief) in the 1820s and 1830s. Quite how the man born at No. 6 Merrion Street attained such greatness would have been beyond his countess mother Anne, who despaired that her listless and aimless offspring would never make anything of himself. When he was in his early twenties, Anne confessed: 'I don't know what I shall do with my awkward son Arthur.'

But awkward Arthur turned a new leaf. He trained up as a crack horseman, joined the British Army in Ireland, got himself elected MP for the rotten borough of Trim, County Meath, and, rejected in love, symbolically burned his cherished violins by way of dedicating himself fully to military glory. And his new-found dedication paid off. Wellesley arrived at Waterloo as England's conquering hero, bedecked with honours and titles including the newly created one of Duke of Wellington conferred on him just the previous year.

While Wellington had been a slow-starter on the march to greatness, his rival Napoleon Bonaparte had sealed his reputation as France's greatest general while still in his twenties. At thirty he installed himself as First Consul, giving himself dictatorial powers. Five years later in 1804 he had himself crowned Emperor of the French en route to becoming the effective ruler of much of the Continent.

But Napoleon bit off more than he could chew. He was ousted from power in 1814 and exiled to the sleepy Mediterranean island of Elba. In February 1815 he escaped, landed in France, and took up where he'd left off as the supreme ruler of France. In June he arrived at Waterloo in modern-day Belgium at the head of an army of some 70,000 to 100,000 men. There, he joined battle with Wellington's larger British Army, which included many thousands of Irish soldiers.

Waterloo is one of the most written-about battles in world history, and a striking number of the eyewitness accounts dwell on the atrocious weather conditions and how they might have had a profound effect on the outcome of the conflict.

The Battle of Waterloo was actually the closing act of a three-day encounter that ebbed and flowed across the rolling landscape of Flanders. Some historians have described as 'Wagnerian' the terrifying foment of thunder and lightning with the flash and crash of cannon fire on the evening before the main encounter. That night one British officer wrote in a letter that, as the thunder and lightning drew closer, 'with it came the enemy ... the rain beating with violence, the guns roaring, repeated bright flashes of lightning attended with tremendous volleys of thunder that shook the very earth ... the night came on, we were wet to the skin ... the bad weather continued the whole of the night ... It would be impossible for any one to form any opinion of what we endured this night. Being close to the enemy we could not use our blankets, the ground was too wet to lie down ... the water ran in streams from the cuffs of our jackets, in short we were as wet as if we had been plunged over head in a river. We had one consolation, we knew that the enemy were in the same plight.'

When the morning of the battle dawned, the enemy French were, in fact, in a worse plight. Circumstance dictated that Napoleon had to go on the offensive, with Wellington in defence mode. The torrential downpour had turned the battlefield into a mudbath,

which would slow down the movement of men, horses, and particularly heavy artillery, on both sides. Embedded in their defensive positions however, Wellington's troops didn't need to move nearly as much as their enemy.

Many historians believe that Napoleon's fatal mistake at Waterloo was to take bad advice from his generals and delay his assault for vital hours in the hope that the ground would dry out enough to improve the mobility of his troops. Confounding those hopes, the weather remained dank and the delayed attack went ahead in conditions that weighed heavily against the advancing French. Various accounts of the battle have given a range of examples of how the weather intervened.

The French artillery kept getting stuck in its tracks, and where big guns could fire freely their cannonballs got plugged in the soft mud rather than having the usual deadly impact of skimming over the ground, wrecking life and limb. The sticky ground exhausted the horses of Napoleon's normally formidable cavalry before they'd reached the enemy lines, with some charges taking place at a canter instead of a gallop. The thick clouds of smoke from the cannon and musket fire mingled with the dank, overcast, misty conditions to cloak the battleground in a dense fog where troops found it hard to make out either friend or foe. Some historians have accused Napoleon of neglecting his communications network before the battle, with the result that key dispatches got lost in the murk.

Other experts have concluded that the weather had another cruel twist in store for the French infantry, who couldn't keep their powder dry as they waded towards the enemy lines through shoulder-high fields of sopping-wet rye. The rye, it is thought, would have dampened the gunpowder in their muskets and pouches, leading to a high rate of misfires when they were staring into the whites of the eyes of their British opposites.

In the end, the adverse weather conditions coerced Napoleon into holding back his dawn attack until the early afternoon. Many military strategists and historians have argued that this weather delay was Bonaparte's fatal mistake, allowing a large Prussian force under Field Marshal Gebhard von Blücher to catch up with the action and deliver the crushing blow against the French. Others argue that Wellington had already turned the tide of battle by the time the Prussians arrived to seal the victory. Both schools of thought are agreed on the reflection of a deeply relieved Wellington that it was all too close for comfort.

Defeat at the Battle of Waterloo ended Napoleon's so-called hundred days' reign as the Comeback Emperor. Having escaped from nearby Elba, he was now imprisoned half a world away

A TO Z
OF IRISH WEATHER

Pet Day The term Pet Day is common to the Irish and the Scots. It has several glosses. It can be taken at face value as a single fine day in the middle of a bad spell. However, because it is so unexpectedly pleasant, it is often regarded as an act of deceit by the elements, and taken as a warning that the worst of the weather is still to come.

Porridge In County Donegal during the icy January of 1973 the Donegal hotel lobby flexed their muscle having taken offence at a radio advert for Ranks Porridge Oats. The hoteliers complained that the commercial implied that visitors to Donegal would need to get a warm breakfast inside them each morning because the weather in the county was so rank. Ranks withdrew the advert.

Queem A word found in Ulster-Scots to describe conditions that are pleasantly calm or smooth.

on the desolate South Atlantic outcrop of Saint Helena. He was accompanied by his personal physician, Irishman Barry O'Meara, who sent letters to a contact in the London Admiralty claiming that the health of the planet's most famous prisoner was being severely damaged by ill treatment. Napoleon died in 1821 after six years of captivity. His autopsy said the cause of death was stomach cancer, but a thriving cottage industry has sprung up in recent years around the conspiracy theory that he was slowly poisoned by his captors.

For the first Duke of Wellington, on the other hand, the Battle of Waterloo merely set the scene for almost four more decades of power, glory and acclaim. He became Britain's Prime Minister in

Rainbow Chasers Since the rainbow is an optical illusion, there is no end of the rainbow where it is supposed to meet the ground. Hence daydreamers, utopians and people who are plain bonkers are sometimes called rainbow chasers.

Rain Check An Americanism that has slipped into Irish usage through films, TV and song, this is shorthand for postponing an invitation to do something, as in: 'I'll take a rain check on that.' The term comes from baseball in the 1880s where paying spectators were given a cheque entitling them to free admission should the game be rained off. The misspelling of 'check' has stuck.

Red Rain In November 1979, during an unseasonal spell of warm weather, people woke up all over Ireland to discover that their cars and windowpanes had turned red overnight, heavily sprinkled with a fine ruby dust. Scientists at University College Cork issued a statement reassuring the nation that the red dust was harmless. This curiosity happens on rare occasions when the air currents line up perfectly to convey red dust from Saharan sandstorms all the way to Ireland. To reach the airspace over Ireland the sand will normally be carried north and west by an anticyclone, before being washed down as red rain by a downpour.

 1828, but by then the name Wellington was well on the way to becoming one of the gold-plated brands of the era. In 1840, 150 British and Irish settlers landed at the mouth of the Hutt River in New Zealand. Shortly after, they called their settlement Wellington, and in 1865 the town named for the Dubliner became the capital of New Zealand. The popular dish Beef Wellington is also named for the Duke, along with variations such as Sausage Wellington, Lamb Wellington and Salmon Wellington. There are several theories as to why the pie dish was given the name Wellington. The most persuasive is that the pastry casing around the joint of meat looked like a leg in one of the shiny boots named after the Duke.

For 200 years the people of Ireland, and of other rain-soaked lands, have pulled on wellies as the best protective footwear known to mankind against muck, rain and the worst the elements can throw at us. Legend has it that a year or so before Waterloo, Arthur Wellesley instructed his London shoemaker to modify the popular Hessian boot which was designed to slip smoothly into a horse's stirrups and was standard issue for Europe's military officer class.

Wellington was delighted with his modified boot which was fabricated in soft calfskin leather and was cut for a skin-tight fit around the leg. It was robust enough for battle, yet comfortable for the endless round of glamorous balls the Duke attended. A hit 1815 portrait by the artist James Lonsdale – which showcased Wellington posing rampant in his new footwear on the field of Waterloo – promoted Wellington Boots as the fashion must-have of the day amongst gentlemen who wished to emulate their hero.

But, as everyone knows, the real attraction of the wellington boot as it has come down to us today is the sturdy protection it provides against the elements. In 1852 the American inventor Charles Goodyear introduced his new rubber vulcanisation process to Europe and by the following year the waterproof rubber wellie was replacing the traditional clogs of the Continent's field workers, first in France and then across Northern Europe.

In the year that Goodyear arrived in Europe with the technology that would make the wellington boot one of the world's greatest weather-beaters, Arthur Wellesley died at the great age (for the time) of eighty-four years. Strangely, to a generation steeped in today's celebrity culture, the passing of the world-famous Dubliner didn't even make the lead story of a top newspaper like *The Manchester Times*, which instead led with news of cotton shipments due to arrive from America. Somewhat bizarrely too to the modern reader, *The Morning Chronicle* expressed upset that the 84-year-old had been denied 'the highest form of a soldier's death – to be cut off in the very hour of victory and triumph' on the battlefield.

Throughout his life Wellington had played down his Irishness, allegedly remarking that being born in a stable doesn't make one a horse. Many doubt he said it, but *The Morning Chronicle*, like the rest, hailed him as 'the greatest English hero'.

He wouldn't have wanted it put any other way.

DEATH BLOWING IN THE WIND

Potato Blight and the Great Famine

AROUND THE YEAR 1810 a new breed of potato arrived in Ireland from the Americas. Called the Lumper, it positively flourished in the damp weather and heavy soil of the west of Ireland. By then, up to one third of the Irish populace was getting by on a diet of potatoes for breakfast, dinner and tea, with a little butter or milk added as a treat.

While many of Europe's peasants were mired in a miserable daily struggle against starvation, the Irish peasantry was thriving to the point that the population of the island had soared from around 2 million to 8 million in under two centuries. The potato fuelled this remarkable growth spurt. To this day, this endlessly versatile tuber is the only vegetable on the planet that provides all of life's essentials in one handy pack. A true superfood, it contains carbohydrate, fat, proteins, Vitamins B and C, and a range of minerals including iron, calcium, potassium, sodium, magnesium and zinc.

The Irish climate was a key factor in the rapid and widespread take-up of the potato by the Irish people. The mild, damp weather had always favoured pasture (the grazing of cattle and sheep) over tillage (the cultivation of crops such as wheat and oats), especially in the more rainswept west. The adaptable potato grew well on the more waterlogged soils west of the Shannon where anyone with access to a small patch of land could use the 'lazy bed' planting method. Cultivation at its most basic, lazy beds were shallow furrows overlaid with turf, and seaweed or animal fertiliser. It was simple but effective.

But while the Irish climate made an inviting home for the potato, it also enabled the spread of the fungal infestation *Phytophthora infestans*, or potato blight. The blight prospered in the mild, damp conditions which encouraged it to reproduce by casting its spores on the winds that were heavy with moisture. The fungus first attacks the leaves and the stalk, before spreading down beneath the surface to consume the edible tuber.

The Irish summer of 1845 was muggy and sunless. There was nothing unusual in that. But the strain of *Phytophthora infestans* that arrived that year was new and highly contagious. The botanical epidemic was first observed along the eastern seaboard of the United States in 1843. In the space of just two years it was wreaking havoc with crops as far afield as the American Midwest. By the early summer of 1845 it was raising worries in the big potato-eating pockets of Europe, starting in the Low Countries. Over the next few months it turned up in Ireland, France, Germany, Switzerland and Scandinavia.

Historians now believe that the blight was able to make the rapid leap from America to diverse parts of Europe because it was carried with seed potatoes, and in the seabird manure called guano, imported by the United States from Peru. Ironically, the importation of new seed potatoes was intended to freshen up the crop with new genes, keeping it robust and immune from infection.

The blight was first seen in Ireland in early September 1845, but it was not until October, when the main crop was to be lifted from the ground, that the real damage was uncovered. As November approached, *The Spectator* gravely noted: 'Ireland is threatened with a thing that is read of in history, and in distant countries, but scarcely in our own land and time – a famine. Whole fields of the root have rotted in the ground, and many a family sees its sole provision for the year destroyed.'

The weather intervened to spare some parts of Ireland the full impact of the blight in the autumn of 1845, as early frosts stopped the spores in their tracks. At the end of that first year of blight, approximately one third of the potato crop had been destroyed but full-scale famine had been averted.

There was a widespread confidence that such a vicious blight could not strike two years in a row, and in the early part of 1846 the seed that had been salvaged and stored over the winter was planted. The new plants dug up in the spring of 1846 appeared to be healthy and the crisis seemed to have passed. Filled with a new optimism, the poorest classes kept with custom and rented small plots of 'conacre' land to plant with potatoes.

That optimism was quickly dashed. The blight had hit the Low Countries far harder than Ireland in 1845, with parts of modern Belgium and The Netherlands losing between 65 per cent and 85 per cent of their crop. Now, however, it reappeared in the west of Ireland, incubated and spread by the mild, moist Atlantic breezes. It moved pitilessly eastwards with the wind, withering crops across Ireland and the Scottish Highlands. An uncommonly dry summer in England and across the Continent kept it at bay elsewhere that year.

The horror and tragedy of the Great Famine has been documented in minute detail elsewhere, so here we can stick to the effects of the climate. The blight relented in 1847 and the crop did not fail, and yet this year has been highlighted in Irish folk memory as Black '47. The lack of blight persuaded many in authority that the worst had passed and that normal service would be resumed quickly. But the authorities hadn't counted on the fact that an entire economic system based precariously on the potato was in a state of collapse, and that since very few seeds had been saved for planting, the potato crop, while healthy, was tiny.

Black '47 is so-called because it saw a great part of Ireland convulsed with mass starvation, mass evictions and mass emigration. It was horror without respite. The blight returned in the summer of 1848, and again on a lesser scale in 1849 and 1850. In the midst of it all, Britain's Lord Lieutenant of Ireland pitied the starving poor whose 'constitutions are so broken that they are more than half rotten from all they have gone through'. They were, he remarked, 'feeding more on hope than on meal', and with both hope and food gone they were condemned to 'now die in swarms'.

And that they did.

FORCE 10 AT FASTNET

Dangerous Waters and the
Father of Seismology

'I REMEMBER ABSOLUTELY VIVIDLY one huge wave that hadn't broken. There was lots of phosphorescence around that night and this wave was a blackness in the shimmery green of the broken waves. It seemed to come at ninety degrees to the rest of the wavetrain. I tried to turn downwind but we didn't have enough boatspeed.'

That was the eyewitness testimony of Stuart Quarrie, a survivor of the worst killer storm in the history of yacht racing. In 1979 Quarrie was a competitor in the biennial Fastnet Race, a seafaring sprint over 608 nautical miles from Cowes on the Isle of Wight, around the Fastnet Rock south of Cork, and back to Plymouth on the English mainland.

Recalling that huge, black wall of water thirty years later in 2009, Quarrie continued: 'It hit us nearly abeam. That's the last I remember of that wave because I was washed out of the boat. My harness clip opened up – the hook got caught sideways in the D-ring and came undone and I went swimming.'

Stuart Quarrie lived to tell the tale. Fifteen of his fellow sailors perished in the disaster. So too did three of the 4,000 rescuers mobilised in the greatest maritime life-saving operation ever mounted in peacetime. Quarrie recalled: 'This had all happened within a couple of hours of the midnight forecast. The scariest thing was how quickly it all developed.'

The BBC shipping forecast had predicted choppy weather off the south of Ireland for that August day, but the very worst the crews anticipated was a Force 8 gale. According to the Beaufort Wind Scale developed by Irishman Francis Beaufort, a Force 8 Gale at sea would equate to: 'Moderately high (18–25 ft) waves of greater length, edges of crests begin to break into spindrift, foam blown in streaks.' On land a Gale 8 would manifest itself as: 'Twigs breaking off trees. Generally impedes progress.'

Instead, what engulfed the yachts out of the blue was a Force 10 that had sped across the Atlantic, picking up velocity and ferocity as it travelled. A Force 10 is a very different beast from a Force 8, and is described in Beaufort's reckoning as: 'Very high waves (29–41 feet) with overhanging crests, sea white with densely blown foam, heavy rolling, lowered visibility.' On land, this would translate as: 'Trees broken or uprooted. Considerable structural damage.'

In 1985 the Fastnet Race once again made global headlines when singer Simon Le Bon of one of the world's biggest bands of the time, Duran Duran, needed rescue from his capsized yacht. In contrast, a giant rogue wave so high that it reportedly washed over the Fastnet Lighthouse didn't rate a single column inch in the newspapers that year.

It was many years later that the retired lighthouse keeper Dick O'Driscoll told a journalist from *The Economist* of a wave that must have towered 40 metres in height. According to the magazine's report:'In stormy winter weather the "big seas would come sailing up over the entire building like the field of horses in the Grand National", as one former Fastnet keeper put it. Sometimes, there were almost disastrous consequences. Mr O'Driscoll remembers a storm in 1985 when a wave reached as high as the light and came crashing through the glass, overturning the vat of mercury and sending the poisonous liquid pouring down the stairs. He doubts the tower would have withstood another wallop as great as that, but it never came. Suddenly, there was a great calmness.'

While the Irish newspapers of 1985 are silent on this massive wave, it has taken on a life of its own as an internet 'fact', with Wikipedia and several other sources giving its height at a very precise 48 metres or 157 feet. Make of that what you will.

The Fastnet Lighthouse itself was born out of an earlier shipping disaster caused by bad weather. In 1847 the *Stephen Whitney*, an American sailing packet carrying cotton, corn, clocks and cheese, emerged from three days of dense fog and sailed straight onto the rocky shore of West Calf Island off Fastnet. Of the 110 souls on board, 92 died. The authorities decided that a lighthouse must be built on Fastnet to prevent further tragedies.

The ironwork for the new lighthouse was supplied by the Dublin iron foundry firm R. & J. Mallet, which was being run by John Mallet and his precocious son Robert. At the tender age of sixteen Robert entered Trinity College Dublin to study science and mathematics. To this day the iron railings surrounding the famous university bear the mark of the foundry that made them, R. & J. Mallet. Robert, however, would go on to make a far greater mark in the field of learning.

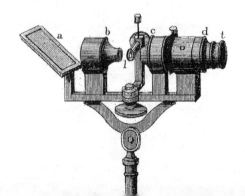

IRISH WEATHER PROVERBS
– WARNING SIGNS

Bad weather is like a clattering train.

Snow is due when the cat washes behind both ears.

When soot falls down the chimney the days ahead will be dismal.

It will rain when the turf flame is blue in the fire.

A spring day has many moods and fits.

When the sun has thin, straggling legs, bad weather lies ahead.

It will be rough tomorrow if the sea is full of foam at full tide.

If there is good visibility at sea, bad weather is due.

The old crone's little finger gets itchy when damp is coming.

When foam makes white roads in the sea, bad weather will come.

Frost is more treacherous than snow.

A good fall of snow is worth seven years of manure.

A windy day is not for thatching.

On 9 February 1846 he presented the Royal Irish Academy with his paper *On the Dynamics of Earthquakes*, which is one of the foundations of modern seismology. Indeed, Robert Mallet is credited with coining the terms 'seismology' and 'epicentre'. In October 1849 he began a series of experiments on how sound or energy moves through sand and rock. Together with his son, a Trinity College geology student, he carried out a landmark experiment on Killiney Beach in south Dublin.

In an attempt to prove that energy moves through sand and rock in waves that can be measured, the father-and-son team designed a controlled experiment to test their hunch. They buried a keg of gunpowder on the strand and detonated it. Using a seismoscope of a type invented by the Chinese two millennia earlier, they measured the 'elastic' wave that travelled through the sand from a distance of half a mile. Mallet's report on his Killiney Beach experiment is considered *the* foundation paper of the modern study of earthquakes.

In 1857 the Italian city of Padua and its hinterland were devastated by a huge earthquake which claimed the lives of some 20,000 people. The quake was the third biggest in recorded history and Mallet made a beeline for the region, with the support of two eminent sponsors Charles Lyle and Charles Darwin who helped him win a funding grant of £150 from the Royal Society of London.

At the scene of the devastation Mallet diligently recorded the damage with the help of a French photographer. The use of photography as a means of analysis was revolutionary, as Mallet employed it to document the way buildings were cracked, walls toppled, and soft ground fissured. Mallet believed that an earthquake consisted primarily of a compression followed by a dilatation. In other words, by observing the way walls and pillars cracked and fell, he traced a path in reverse to the original epicentre of the earthquake. As it turned out, some of his assumptions were wrong, but he had kick-started modern seismology and set it probing in the right direction.

The Dubliner's report was presented to the Royal Society of London and was warmly welcomed as a seminal scientific work. In 1862 he expanded his thinking in a follow-up paper, *The Great Neapolitan Earthquake of 1857: The First Principles of Observational Seismology in two volumes*. In it, he put forward evidence to show that the epicentre of the Neapolitan earthquake was around nine geographical miles below the Earth's surface.

Blind for the last years of his life, Mallet died in London in 1881. His peers marked his passing as that of a true scientific great.

A TO Z
OF IRISH WEATHER

Sea Area Forecast The sea area or shipping forecast broadcast by Met Éireann is quite parochial, in that it refers to a slim zone around the Irish seaboard extending 50km/30 miles out to sea. The shipping forecast broadcast daily by the BBC incorporates Irish maritime regions into a bigger picture, as Britain is charged under international agreements with monitoring the weather conditions around the British Isles and far beyond, both north and south. The British system divides up the surrounding part of the North Atlantic into a grid containing sea areas named after headlands, lighthouses, and the mouths of major rivers. These sea areas extend to the coasts of Denmark and Norway in the east (Fisher, German Bight, etc.), to South East Iceland in the North, and to Trafalgar off Portugal in the South. The Irish sea areas in this system include Shannon, Rockall and Fastnet.

Snowballs From Hell We tend to think that the nasty practice of concealing a stone in a snowball is a modern deviance thought up by the type of young gurriers who drop heavy objects from motorway footbridges and shoot airguns at pigeons. Not so. In 1786 *Faulkner's Dublin Journal* fumed: 'The pernicious custom of throwing snowballs has arrived at an intolerable height. No less than a dozen decent persons have been desperately wounded by stones and brickbats wrapped up in these missile weapons of barbarous amusement.'

The following year one citizen was pushed over the edge by the practice. *The Dublin Chronicle* reported: 'A gentleman passing through Marybone Lane was hit by a fellow in the face with a large snowball, upon which he immediately pulled out a pistol, pursued the man, and shot him dead.' The *Chronicle* thoroughly approved of what it considered to be an open-and-shut case of justifiable homicide. It warned would-be troublemakers: 'These deluded persons are therefore cautioned against such practices, as in similar circumstances they are liable, by Act of Parliament, to be shot without any prosecution or damage accruing to the person who should fire.'

Solstice A solstice occurs twice a year, as the sun appears from the Earth to reach its highest altitude (June in Ireland) and its lowest (December) above the horizon. The word compacts the Latin '*sol*' for sun, and 'sistere', meaning to stand still. This is because, as seen from the Earth, the sun appears to stand still in the sky before reversing its direction.

Stokes' Law Born in Skreen, County Sligo, George Stokes (1819–1903) came up with a law of physics to express the drag force acting on spherical objects such as raindrops. His law explained how small water droplets or ice crystals can remain suspended in clouds until they grow to a critical mass and fall as rain, snow or hail. Stokes left behind a vast body of pioneering scientific work, and his insights are credited with providing the foundations for three Nobel Prizes awarded to others.

Sunspots In the 1870s the leading British economist William Stanley Jevons posited a link between sunspot activity and economic crashes. He argued that sunspots influence the Earth's weather which affects crop growth which, in turn, impacts on the economy. The study of magnetic sunspots, which produce temporary regions of lower heat on the sun's surface, is in its infancy. Scientists have identified an eleven-year cycle of sunspot activity, and are attempting to come to an understanding of the effects these have on the Earth.

Sun Worship, Ancient The sun was worshipped by the ancient Irish, although it was just one of many objects of adoration in the natural world. In his *Confession*, Saint Patrick admired the sun as an object of great splendour, but he warned the Irish that it was not in itself a god, but was at the command of the one true Christian deity. He wrote: 'This sun which we see rises daily for us because He commands so, but it will never reign, nor will its splendour last. What is more, those wretches who adore it will be miserably punished. Not so we, who believe in, and worship, the true sun, Christ, who will never perish.'

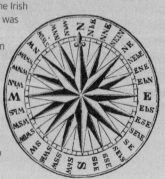

Sun Worship, Modern The suntan has gone in and out of fashion over the centuries. At the beginning of the twentieth century a tanned skin was regarded as the mark of a low station in life, as it was the unhappy lot of the labouring classes to have to work exposed to the elements in the great outdoors. Ladies of refinement went to considerable lengths to keep themselves pale and interesting, to the extent of using arsenic, lead and other potentially harmful substances to bleach their skin.

The attitude of the ruling classes to the sun began to change in 1903 when Faroe Islander Niels Finsen was awarded the Nobel Prize for Medicine for his Finsen Light Therapy. Finsen's treatment proved a miracle cure for the scourge of rickets by correcting a Vitamin D deficiency in sufferers by exposure to sunlight. The style icon Coco Chanel is credited with giving sunbathing a huge boost in the 1920s. Legend has it that she accidentally got sunburnt, and arrived back in Paris with an unintentional suntan, setting a trend that has been with us ever since.

By the 1930s *The Irish Times* was carrying adverts for summer sun cruises. One report from 1935 recommended taking a cruise to North Africa, Madeira and the Canaries with Canadian Pacific Sun Tan Cruises, although the writer did admit that the all-in cost of £12 for the thirteen-day trip was not for every pocket. The same newspaper in the same year offered the following sun protection advice to those in search of a bronzed skin: 'If a suntanned complexion is desired, here is a simple way of producing an even tint without painful burning. Sponge the skin lightly with malt vinegar and allow it to dry on before going out on a sunny day.'

Twenty years later, in 1955, *The Irish Times* reflected on the fact that the suntan had performed a 180-degree somersault in the space of half-a-century, going from a signifier of the lowest social standing to the highest. The paper noted: 'There is as much social snobbery connected with a suntan as with the ownership of motorcars or the acquisition of titles. At the bottom of the scale comes the paleface, who has apparently never ventured outside the door except during the hours of darkness; at the top is the polished mahogany surface which has taken months of careful oiling and exposure to bring to full bloom ... One must not discard the theory that Hollywood – or more precisely Technicolor – has played its part. If the screen were not peopled with strong sunburnt men, the streets of this and other cities might not at the moment be full of weaker, and redder, faces.'

THE WORLD'S FIRST WEATHER FORECAST (GETS IT RIGHT!)

Robert FitzRoy Sets Up
The First Met Service

THE FATHER OF THE WEATHER FORECAST as we know it today was Robert FitzRoy, the most distinguished protégé of County Meath's great weather pioneer Sir Francis Beaufort. The Irishman helped the young FitzRoy secure the captaincy of HMS *Beagle*, which carried Charles Darwin on the 1831 voyage of discovery that inspired and informed Darwin's world-shaking *On the Origin of Species*.

Fearing a backlash of Biblical proportions from a God-fearing Victorian society, Darwin put off publishing his theory of evolution until 1859. That same year a maritime disaster in the Irish Sea pushed FitzRoy towards making his second mark on posterity. During a turbulent two-week period in the winter of 1859, some 200 vessels were sunk by storms in the waters around the British Isles.

One of them was the *Royal Charter,* a steamship near the end of a two-month voyage from Australia carrying 430 passengers and a large cargo of gold bullion. Queenstown (now Cobh) in County Cork was the ship's last quick port of call before casting off for its final destination of Liverpool.

Just a few hours out of Cobh, the *Royal Charter* was smashed against the rocks of Anglesey off the coast of Wales with the loss of 400 lives. Observers at the British Admiralty were able to piece together the course of the storm that wrecked the *Royal Charter* by joining the dots between observations from the few stations gathering data in Ireland and Britain. The public outcry surrounding the big loss of life nudged the Navy into designing and building a system of storm warnings for ships going to sea. As the head of Britain's fledgling met office – his full title was Meteorological Statist to the Board of Trade – Robert FitzRoy was charged with the task.

Soon to become a Vice Admiral, the energetic FitzRoy's former jobs had included being Member of Parliament for Durham and Governor of New Zealand. An accomplished weatherman, he'd designed and invented several types of barometers, and in 1863 he would publish *The Weather Book* which was instantly acclaimed as a groundbreaking addition to the field.

A born organiser, FitzRoy quickly put in place a network of weather stations around the coastline of the British Isles. Each was equipped with the recently invented electric telegraph, which enabled weather reports to be sent without delay to a central office where the data could be correlated and storm warnings telegraphed to the harbourmasters of the chief ports of Britain and Ireland.

With the invention of wireless still some decades off, the telegraphed warnings had to be passed on to ships within sight by sending visual signals from the shore. These signals were based on the traditional semaphore system. A northerly gale was indicated by a cone-shaped flag pointing upwards. A flag in the shape of a cylinder told sailors to expect gusts from any or all directions, and so on. The harbourmaster's job was to have the appropriate flags erected within thirty minutes of receiving the telegraph message.

With forty stations providing data in real time by the close of 1860, FitzRoy was now in a position to compile detailed weather charts, which were being constantly updated. This in turn opened up the tantalising prospect of being able to make weather predictions with a fair degree of accuracy.

Attempts to predict the weather go back far into the mists of time. By around 650 BC the Babylonians, who created a great civilisation in what is today Iraq, were using an unscientific mix of cloud patterns and astrology to make forecasts. The ancient Greeks, Chinese and Indians also augmented their observations of nature with a generous dash of hocus-pocus.

While some experts claim that the word 'forecast' was first used in the context of weather predictions almost two centuries before FitzRoy's time, others credit him with coining the term 'weather forecast' with very deliberate precision. He felt that the word 'forecast' had a suitably scientific ring to ward off any connotations of prophecy or supernatural help.

To show confidence in his network of weather stations, FitzRoy provided *The Times* of London with the world's first weather forecast published for public consumption. His forecast for Britain, for 1 August 1861, was brief and extremely skimpy on detail. It simply said:

North – Moderate westerly wind; fine.
West – Moderate south-westerly; fine.
South – Fresh westerly; fine.

Short and sparse it may have been, but in terms of accuracy it couldn't have been finer. Getting it right first time out was a slice of beginner's luck that 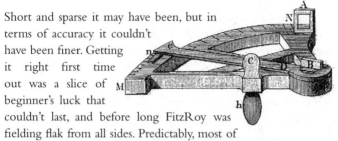 couldn't last, and before long FitzRoy was fielding flak from all sides. Predictably, most of the complaints were directed at inaccurate forecasts, but the Vice Admiral had also made an enemy of a rich and powerful lobby group who were out for his head. The owners of the fishing fleets of Britain and Ireland didn't just object to incorrect weather forecasts, they were furious that FitzRoy was forecasting bad weather at all! The owners claimed, correctly, that storm warnings from FitzRoy's forecasters were costing them money as their fishermen refused to take their boats out to sea.

Another powerful group that poured scorn and bad publicity on FitzRoy's forecasts were the publishers of the hugely popular almanacs, the leisure and infotainment magazines of the day. The most famous of these, *Old Moore's Almanack*, had been publishing long-range weather forecasts since it first appeared in 1697, basing its predictions on astronomy, astrology and blind guesswork. FitzRoy's scientific forecasts were a threat to one of the almanacs' big selling points, and the publishers wasted no opportunity to discredit him and his methods.

The constant criticism piled the pressure on a man who began to exhibit the classic symptoms of clinical depression. His fragile state of mind was compounded by failing physical health and the worries that he was spiralling towards bankruptcy.

When Robert FitzRoy took his own life in 1865 the fishing fleet owners didn't spend too much time in mourning. With their sworn enemy gone, they succeeded in browbeating the British government into abandoning the Vice Admiral's system of scientific weather forecasts. But FitzRoy had his vindication from beyond the grave. After a short absence, the weather forecasts were restored by public demand of the fishermen who had seen their true worth in saving lives. Robert FitzRoy had become nothing less than a hero to the fishing communities of these islands. As the wife of a Scottish fisherman reportedly mourned upon learning of his death: 'Who will look after our men now?'

A TO Z
OF IRISH WEATHER

Trade Winds The Trade Winds in the Northern Hemisphere blow east to west from North Africa to the Caribbean, and have their opposite number in the westerlies or anti-trades, which blow onto Ireland's western seaboard from the Atlantic. Contrary to popular belief, the name Trade Winds has nothing to do with the fact that these winds were vital to transatlantic commerce in the age of sail. They were labelled Trade Winds much earlier, deriving their name from the Middle English word *trade*, meaning 'course', 'track' or 'path'.

Truancy When the Dáil debated a new School Attendance Bill in 1942 there were calls to punish the parents of children who failed to show up for lessons on a regular basis. However, one opposition deputy, James Dillon, argued that there was little point in blaming the parents of stay-away kids when the country's schoolhouses were not fit for purpose when it came to keeping out the weather. Dillon fumed: 'I know schools in rural Ireland, never mind schools in the City of Dublin, to which I should be very slow to send a child. They are dirty, damp, draughty, miserable hovels, with sanitary accommodation substantially inferior to the ubiquitous ditch. Have we put that state of affairs right before we determine to tighten the screw in order to force children to go to such places? ... I know the wretched little damp, windy, ill-kept bandboxes scattered about the bogs into which children are herded every day, with wet clothes, wet feet and perished with cold. They bring a sod of turf under their arms and try to kindle a bit of fire, but long before the fire is lighted and is giving out any heat, it is time for the children to go home.'

The Ceann Comhairle shot down Dillon's impassioned plea for weatherproof schools, ruling: 'The Deputy has now gone outside the ambit of the Bill.'

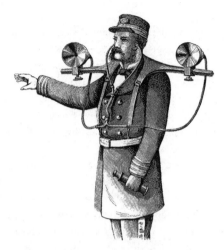

WEATHER AND WIRELESS

Ireland's Own Guglielmo Marconi Bounces Signals Off The Atmosphere

IN 1858, AFTER SEVERAL COSTLY FAILED ATTEMPTS, the first transatlantic telegraph message was sent from Valentia Island in County Kerry on the westernmost tip of Europe, to Heart's Content on the east coast of Newfoundland. The heavy cable slashed the time to send a message across the great ocean from the ten days it took by ship, but the first ninety-nine word greeting from Queen Victoria still took almost seventeen hours to reach the other end. As the technology improved, so did the speed, but by the closing years of Victoria's long reign the race was on to send electrical signals higher, faster and stronger.

It is widely accepted that the clear winner of that race was Guglielmo Marconi, a pioneering genius whose Italian name often obscures the fact that he was born to an Irish mother and married an Irish wife. In 1914 Valentia Island became home to a state-of-the-art Marconi radio station and within a decade it was the busiest marine wireless station in the British Isles, sending out storm warnings to ships at sea, taking in distress calls and alerting the rescue services to vessels in trouble.

In all of history just a tiny few very special individuals have profoundly changed the world about them. Guglielmo Marconi was one of them. When he died in July 1937, feuding nations paused to salute the passing of the chief architect of what would become today's global village. A visionary of genius, a shrewd businessman and a hopeless show-off, Marconi held centre stage with a command that would have relegated showmen like the late Steve Jobs or Richard Branson to the role of mere street buskers.

Marconi's mother was Annie Jameson of the famous whiskey dynasty, and the young Guglielmo romped the grounds of the family's stately home at Montrose in Dublin. The white-walled mansion still stands on the land that, appropriately, now houses the national broadcaster RTÉ.

A school dunce, Guglielmo gravely disappointed his severe Italian father, but his mother spotted his obvious scientific flair and encouraged him to muck about with wires and tubes in the attic, although she would scarcely have approved of his following in the family distilling business with the covert poitín still he built there.

As luck would have it, the youngster's lecturer and neighbour Augusto Righi was a pioneer of radio-wave experimentation and invented an oscillator to boost their transmission. Trouble was, nobody could think of any practical use for them. The teenage Marconi made that giant imaginative leap. His brainwave was to develop a telegraph system that used airborne waves rather than wires. It was a brilliant insight and

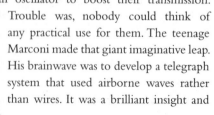

its timing was perfect, because the existing telegraph system was slow, cumbersome and always liable to short out in the wet or blow over in the wind.

The twenty-year-old Guglielmo cracked the problem in 1894. Still in his attic, he succeeded in making an electric buzzer ring by sending a radio signal from a transmitter yards away. He dragged his mother from her bed into his makeshift workshop and showed her the future. The next day he astonished his father, who emptied his wallet on the spot to pay for the materials to advance the experiment. Guglielmo moved his work from his attic to his garden, and within a year was ringing buzzers over a distance of eight miles.

The Italian government wouldn't touch Marconi's seemingly supernatural device with a telegraph pole, and the *Belfast Newsletter* refused to believe his boasts that a signal could pass through 'every obstacle of houses, rocks and mountains'. It scoffed: 'In what way would a message be transmitted to London? Is Marconi a master of telepathy as well as telegraphy?'

The Royal Navy swept aside such doubts, and in 1896 gave Marconi a contract, although landing the deal wasn't plain sailing. Italian 'anarchists' were the Al Qaeda of the day and customs officials tore his suspicious 'black box' to pieces. Brimming with a sense of his own manifest destiny, he rebuilt his device and delivered a killer sales pitch to the navy, who immediately saw its potential to revolutionise land/sea communications.

Back in Ireland, the showman in Marconi was emerging. He astounded the citizens of Dublin with a display from the clock tower of Rathmines town hall, and those of Ballycastle in Antrim by sending a message from Rathlin Island six miles out. In 1898 he was back home in Ireland, where he combined the twin loves of his life – sailing and showing off – at the Kingstown (Dún Laoghaire) Yacht Regatta. *The Dublin Daily Express* commissioned him to send wireless reports of the races and the weather conditions. It was another historical first for Marconi – the first ever press report by wireless. *The Times* of London recognised the significance, saying it proved 'the commercial value of the Marconi system'. Impressed, Queen Victoria requested a wireless set for her royal yacht.

In 1901, Marconi built a wireless transmitting station at Marconi House, Rosslare Strand in County Wexford. Using the station to boost a signal from Poldhu in Cornwall to Clifden in County Galway and then onward, he claimed to have sent the first transatlantic message in Morse Code from Britain to Newfoundland. Not everyone believed him, but when he backed up his claims with further transmissions in 1902 the whole world sat up and took notice.

In order for his signals to travel some 2,000 miles from Ireland to the coast of Canada, they needed to bounce twice off the newly discovered ionosphere. The ionosphere exists in the upper part of the Earth's atmosphere where it influences, and is influenced by, the weather closer to the surface in ways that are still not fully understood. Marconi and his contemporaries quickly grasped that the magnetic push and pull of the upper atmosphere caused what's loosely known as atmospherics.

Readers who grew up with the radio and television of the 1970s and earlier will recall vividly how atmospherics could ruin the reception of both media. In the days when RTÉ and the BBC severely rationed pop music on the radio, Radio Luxembourg attracted huge audiences from these islands for its diet of chart hits. But if there were thunderstorms over Europe, or there was an abnormal heatwave, Radio Luxembourg became a painful cacophony of hiss punctuated with loud farts of familiar tunes. Watching TV, by the same token, could resemble staring at a novelty snowstorm in a fish tank depending on what the weather was doing.

The interruptions caused by atmospherics threatened the ambitions of the early wireless companies to wipe out and supplant the telegraph business. So, as they explored and investigated this new weather-related field, their prime concern was to eradicate those pesky things known variously as 'strays', 'static' and 'parasitic signals'.

It seemed self-evident to the early experimenters that the disturbances were in some way related to lightning storms, and in 1895 Russian wireless pioneer Alexander Popov developed a rudimentary radio device for detecting approaching storms well in advance. In the early years of the twentieth century others explored the potential for using radio as a tool of weather forecasting. Marconi's periodical *Marconigraph* kept tabs on these developments, as did *Weather and Wireless Magazine* which announced that from now on radio would play a key role in detecting thunderstorms long before they appeared on the horizon. It said: 'These stray waves have been "captured" and made to serve a useful purpose.'

For Guglielmo Marconi, the speedy mass marketing of radio served the useful purpose of making him a superstar. By the summer of 1912 the Marconi Company was a global giant and its founder was one of the world's richest and most famous men. He was also one of the most admired, having picked up the Nobel Prize for Physics the previous year. His stock with the public was at an all-time high after the sinking of the *Titanic* in the spring of that year. The SOS signals sent from the Marconi Company radio operators on the sinking ship were credited with alerting the *Carpathia* which changed course and saved the lives of hundreds of passengers.

When the *Carpathia* docked in New York, the tireless self-publicist was on hand to board the rescue ship and claim his share of the credit. In June 1912, he gave evidence at the judicial inquiry, where Britain's Postmaster General gushed: 'Those who have been saved, have been saved through one man, Mr Marconi, and his marvellous invention.' By then

he needed all the good publicity he could get, because that same summer endless reports, headlined 'The Marconi Scandal', linked his name to an illegal insider trading scam involving shares in his company. His hands were clean, but the future Prime Minister Lloyd George and other senior British politicians were caught helping themselves to a sneaky dig-out.

Marconi spent the Roaring Twenties swanning around the world on his private yacht, rubbing shoulders with A-list celebrities. After divorcing his aristocratic Irish wife, Beatrice O'Brien, in 1924, the best man at his second wedding in 1927 was Italy's fascist dictator, Benito Mussolini, who dubbed him 'the conqueror of spaces'. Marconi had become an enthusiastic member of his pal's Fascist Party in 1923.

After completing the first radio conversation (as opposed to Morse signal) between Britain and Australia in 1924 he prophesied: 'There will be no distance in the future over which we cannot telephone.' There were more breakthroughs which would relate directly to a better understanding of the Earth's intricate weather systems. By the end of the 1920s he had harnessed the microwave. Pope Pius XI was the first to benefit with a microwave radio link from the Vatican to the Pontiff's summer residence. Today, scientists are not just using microwave technology to measure and forecast the weather, but some believe that microwaves will eventually be harnessed to exert some control over hurricanes.

Marconi demonstrated the principles of radar in 1935, and was working on a practical radar system at the time of his death two years later. Today Met Éireann and every other meteorological service uses radar to forecast approaching fronts. Radar detects objects by transmitting a pulse of radio waves and looking for signals reflected back from the object. By measuring the time taken for the pulse to reach the object and travel back to the radar, the distance can be calculated. By rotating the antenna and sending out a stream of pulses, the radar can build up a picture of objects. Rain, snow and hail all reflect radar waves, giving a measure of the extent and intensity of the approaching front, which is what we see on our TV weather forecast.

When he died in 1937, aged sixty-three, radio stations, including weather stations, around the world observed a two-minute silence for a man who had made their existence possible. The world had lost a colossus.

IRISH WEATHER PROVERBS – INSECTS AND AMPHIBIANS

When flies gather on the water of a well, good weather is coming.

It will be a soft day tomorrow when threads of gossamer web cover the ground.

The frog is yellow for good weather but his coat is dark for bad weather.

Rain will drench the land when the frog comes in the kitchen.

If the bee is busy after sunset the weather will change.

Spiders in the crannies of walls signal sweltering days to come.

BELOW US
THE WAVES

The Race To Fly The Atlantic

I N 1909, JUST SIX YEARS AFTER the Wright brothers had made
the first manned flight of some twelve seconds and 120 feet,
Frenchman Louis Bleriot astounded the world by piloting his
flimsy aircraft across the twenty-one miles of sea between France
and Britain in just thirty seven minutes. In doing so he claimed the
hefty £1,000 prize offered by the *Daily Mail* newspaper for the
first to achieve such a feat.

Four years later, in 1913, the publisher of the *Daily Mail*,
Lord Northcliffe, put up the enormous prize of £10,000 for the
first person or persons to fly non-stop across the Atlantic. The
competition was suspended with the outbreak of the Great War in
1914, but within days of the war's end in November 1918, adverts
appeared declaring the race on again.

The advertisements outlining the rules were a tad confused. In
the first instance, they announced that the prize would 'be awarded
to the aviator who shall first cross the Atlantic in an aeroplane in

flight from any point in the United States, Canada or Newfoundland to any point in Great Britain or Ireland, in 72 consecutive hours'. In the very next sentence it stipulated: 'The flight can be made either way across the Atlantic.'

In reality, any flight across the Atlantic would have to go with the prevailing winds which blew from North America to Ireland. The flight didn't have to be non-stop. The rules allowed for an aircraft to land on the ocean for rest or repairs, although getting airborne again from the rough Atlantic might prove impossible for even the sturdiest flying boat of the time.

The US Navy discovered this for itself in the late spring of 1919 when it launched the first attempt to fly across the Atlantic. With the US Air Force still just a gleam in the eye of a few in the military, the Navy had charge of the operation to send four Curtiss flying boats on a zigzag route which would involve flying northeast to Newfoundland, then southeast to the Azores in the mid-Atlantic, before turning north again to complete the Atlantic crossing in Portugal and then touching down in London for a fanfare befitting a band of conquering heroes.

While there was nothing in rules of the *Daily Mail*'s grand competition to exclude the US Navy's aviators, they would have found themselves disqualified on several counts, not least that they hoped to beat the worst of the North Atlantic weather by taking a southerly route and making land in Portugal when the *Mail* specified a landing in Ireland or Britain.

The US Navy knew that it was in a race with the prize hunters to get into the record books for making the first airborne crossing but, as the official American souvenir book took pains to stress, the Navy was focused on higher goals. It noted that the *Daily Mail* prize pot would 'send half a dozen British machines to Newfoundland to select their fields and build their hangars in the fog and melting snow' in the spring of 1919. However, the US Navy wouldn't be drawn into a base race with aviators who were in it for the money. The official brochure stressed: 'One way, the world knows, was to make the flight a great sporting event.

The other possibility was to keep it, as closely as possible, to a well organised scientific expedition. Obviously the American Navy could not, even if it would, choose sport in preference to science. Dignity and efficiency forbade that.'

No expense or effort was spared to ensure that the four Navy/Curtiss flying boats, nicknamed 'The Nancies', would secure a victory for science over vulgar sport. The official account explained: 'The chief features of the organisation were government weather reports, the extensive use of radio equipment, the assigning of government destroyers to act as tenders, and the establishing of a line of destroyers reaching from Trepassey, Newfoundland, to Lisbon, Portugal.' Twenty-two 'station ships' lit up like Christmas trees were spaced at fifty-mile intervals along the sea route in order to show the way to the Azores.

From the very start, however, 'the behaviour of the weather and fortune was vile'. One of the four aircraft, the NC-4, did survive the battering of the elements to land in Portugal en route to its final destination in London. All told, the journey had taken twenty-three days.

The official commemorative book was rushed out to acclaim the great feat of the Nancies. It gushed: 'It is all over now. It is becoming history! What has it done for the world? It has given the watching nations a spectacle of imperishable gallantry. The bravery of the attempt, the persistence under hardship, the indomitable courage in peril of those involved will long echo about the first trans-oceanic flight.' Pointing out that it had taken Columbus seventy-one days to cross the Atlantic on his first voyage, the writer declared that the Nancies had 'made havoc with time'.

Sadly for the US Navy and the gallant aviators of the Nancies, by the time the souvenir book hit the shelves to proclaim their glory, history had already been rewritten in a boggy field in County Galway.

Just two weeks after the US aviators touched down in London to popular acclaim, two Englishmen took off from Lester's Field in

Newfoundland with their sights set on immortality and the *Daily Mail*'s £10,000 jackpot.

On the afternoon of 14 June there was a break in the bad weather and tailwinds were predicted all the way to Ireland. As spectators gathered, Captain John Alcock and Lieutenant Arthur Brown packed away sandwiches, coffee, chocolate, whiskey and beer into their flimsy craft. In the cockpit with them was a large battery to power their flying suits, which were essentially electric blankets with arms and legs.

Around 3 a. m., the radio operators at the transatlantic Marconi Wireless Station in Clifden began the hopeful wait for a Morse code signal from the flyers. What neither Alcock nor Brown realised was that their on-board transmitter had gone dead three hours into the flight. As the Clifden operators began their vigil, the two flyers were staring death in the face.

Flying into a storm, they nose-dived towards the choppy sea, pulling out by a hair's breadth. Alcock later remarked: 'The salty taste we noted later on our tongues was foam. In any case the altimeter wasn't working at that low height and I think that we were not more than sixteen to twenty feet above the water.'

Having escaped by the skin of their teeth, the men were immediately thrown from the frying pan into the fire, although that's a poor metaphor given that they flew straight into a snowstorm. Snow and ice shrouded their wings, fuselage, struts and even the engines. As they glimpsed the first rays of dawn their lateral controls had iced up, but they continued to fly on a wing and a prayer, knowing that the coast of Ireland couldn't be more than thirty minutes away.

As they hit the warmer clouds rising off the Irish coast, the snow and ice melted and they were overjoyed to find themselves sitting in a puddle formed at the bottom of their cockpit. A few minutes later they sighted land. They had been aiming for Galway city, but were extremely pleased to find themselves circling over the Marconi wireless station where the operators and the villagers came out to wave. The airmen thought the locals were waving them towards a meadow that looked flat and inviting. In fact, the people on the ground were attempting to warn them away from what was a treacherous swamp. The rough landing left the airplane with its nose stuck in the mud, but Alcock and Brown walked away with barely a scratch. They had smashed the two-week-old American record by covering 1,890 miles in a shade under sixteen hours.

With their stunning feat of speed and bravery, Alcock and Brown ushered in what became known as the Golden Age of Flight, which spanned the 1920s to the late 1930s. Ireland's geographical location and its prevailing winds made it a key hub of the world's most glamorous extreme sport, as dashing pilots of both sexes jousted to set new speed and distance records from continent to continent.

Not everyone was happy with Ireland's eminent position, and one patriotic Frenchman didn't even deem a landing on the British Isles as a true crossing of the Atlantic. Raymond Orteig emigrated to New York in 1912. He worked as a bus boy and café manager and eventually acquired two New York hotels which became popular with French airmen stationed in the United States during the Great War.

In May 1919, with the Nancies on course to log the first transatlantic flight, Orteig offered a big prize of $25,000 for the first aviator to travel non-stop from New York to Paris. His twofold purpose was to put France back on the flying map and to kick-start a transatlantic passenger service. As the grandson of the eventual winner would later put it: 'There was an incredible competition to drive the birth of commercial aviation.'

By the mid-1920s Roosevelt Field on Long Island had become the launch base for several well-known aviators with their eye on the prize. However, an unknown young airmail pilot would steal their thunder. On 22 May 1927, *The Spirit of Saint Louis*, piloted by Charles Lindberg, was spotted over Dingle Bay in County Kerry and the waiting world was put on notice by wireless that the 25-year-old was just hours from Paris and the worship of his countrymen as a great American hero.

After Lindberg's wind-assisted solo crossing, the next, much greater, challenge was to cross the Atlantic in the teeth of the prevailing airflow. This challenge was taken up by Ireland's own world-famous aviator, Dublin-born James Fitzmaurice. In 1915, aged seventeen, Fitzmaurice enlisted in the British Army and the following year survived the slaughter of the Battle of the Somme. He rose through the ranks of the RAF and returned to Ireland in 1922 to help set up the Irish Army Corps following the foundation of the new Free State. Within five years he was commander of the Irish Air Corps. He became an international celebrity in 1928 when, together with two Germans, he made the first transatlantic flight from east to west, travelling from Baldonnel Aerodrome in north Dublin to Greenly Island in Canada.

Dubbed 'The Three Musketeers', the trio were given a ticker-tape parade in New York City along a route ten miles long. Other cities in the US and Canada repeated the honour, while in Washington DC President Calvin Coolidge presented them with the Distinguished Flying Cross. The following year, 1929, Fitzmaurice was invited back to New York State for the dedication of the Fitzmaurice Flying Field, a new aerodrome named after him, serving the town of Oyster Bay. A crowd of some 100,000 came out to cheer. On a visit to Germany in 1933 Fitzmaurice was invited by one of his greatest admirers to drop in for a chat. He reported that he was given a warm welcome by Germany's newly installed Chancellor Adolf Hitler.

As aircraft, navigation and communications improved, the world seemed to shrink and record after record fell to the dashing aviators. In 1932 America's sweetheart Amelia Earhart added one of the closing chapters to the heroic Golden Age of Flight when she became the first woman to fly solo across the Atlantic while breaking the speed record set thirteen years earlier by Alcock and Brown. She also put the west of Ireland back on the aviation map of the world.

A TO Z
OF IRISH WEATHER

Umbrella The umbrella for keeping off the rain appears to have evolved from the parasol which was used as a sun-shading device by the people of Nineveh, the capital of ancient Assyria. The umbrella for deflecting the rain appears to have been unknown, or unused, in Europe until the sixteenth century when the Italian word *ombrella* indicated its advent as a fashionable accessory amongst the aristocracy of Italy's city states. Before that, a gentleman's cloak generally served to keep off the rain. In Daniel Defoe's 1719 novel *Robinson Crusoe*, the castaway makes an umbrella in imitation of ones he'd seen in Brazil. He records: 'I covered it with skins, the hair outwards, so that it cast off the rain like a penthouse and kept off the sun.' The umbrella caught on in Ireland and Britain in the latter half of the nineteenth century, when owning one was the mark of the well-heeled and the upwardly mobile. In 1880 the manufacturers Francis Smyth & Son of Grafton Street and Essex Quay in Dublin ran newspaper adverts headlined **CHRISTMAS PRESENTS, UMBRELLAS, UMBRELLAS.** The advert continued: **'[We] have now ready for inspection latest novelties in Ladies' Plain and Twist Silk UMBRELLAS; Ladies' from 4 shillings, Gents' 7s 6d up to 40 guineas.'**

Vane, Weather Vanes were used to tell the direction of the wind at least from the first century BC, but it was Pope Nicholas I in the middle of the ninth century who decreed that the vanes on church steeples should be in the form of cocks. The Pope reasoned that the cock, which had crowed three times as Peter denied Christ, was a suitable emblem of both Christianity, and of Peter, the founder of the Church of Rome.

Volcano Over the course of the millennia volcanic eruptions have plunged the planet into sustained periods of cold and darkness. Scientists have established that Ireland was stricken in 4375 BC, 3195 BC, 2345 BC, AD 207 and AD 540, with distant volcanoes, or supervolcanoes, the chief suspects.

On a blustery May afternoon that year the dwellers of the city of Derry looked up to see a green and gold monoplane spluttering in circles around the walls. Just before 2 p.m. the plane set down in a field four miles outside the famous city walls. Amelia Earhart stepped unsteadily from the crumpled aircraft.

'Have you come far?' asked the worried landowner.

'From Newfoundland,' she smiled.

Swamped by reporters as she recovered from her effort, she played down her double whammy, insisting: 'I just did it for fun.'

The bob-haired Kansas tomboy who fell out of the sky brought a splash of glamour to Derry unlike any the city had ever seen. It was the age of the liberated flapper, when young women shed Victorian taboos by smoking, drinking, wearing their hair short, their skirts shorter, and even donning men's trousers. Earhart and her fellow superstar aviatrix Amy Johnson were the poster girls of the age and both women were consumed by the urge to prove themselves equals in a rigidly chauvinistic man's world.

Earhart had never intended to put the Derry townland of Springfield on the world map when she set out from Newfoundland. Her plan was to land in  Paris, where the world's media and the general public awaited to acclaim her derring-do. In the event, she was lucky to make it to the western tip of Ireland. Shortly after leaving land behind, her altimeter packed up, meaning she had to guess her height from the waves below for the entire crossing. Four hours out to sea, flames began shooting from her exhaust and mechanical parts began to shake loose. By then she had no choice but to press on, snacking on her packed liquid lunch of meat and tomato stew.

Her fuel pipe was leaking dangerously when she spotted the clouds gathered above the Irish coast, followed quickly by fishing trawlers below. Reaching Connemara, she knew it was Ireland 'by the thatched cottages and peat bogs I'd heard about'.

Asked in Derry why she'd chosen to fly alone, she replied: 'Because if there is a man in the machine you can bet your life he wants to take control.' She added: 'I'm glad I was alone because I can do much more hazardous flying when I feel I am not responsible for the life of someone else.'

As the press conference drew to a close, photographer Victor Barton left for the Derry aerodrome where a plane awaited to whisk his photographs to London for the next morning's papers. Sensing bad conditions ahead, pilot Colin Clarke made a stopover at Belfast's airstrip to pick up the latest weather chart. Shortly after leaving Belfast the plane ran into zero-visibility fog and crashed off the Scottish coast, killing both men.

CLOCKS AND COWS

Experiments With Daylight Savings Time

'EARLY TO BED AND EARLY TO RISE, makes a man healthy, wealthy and wise.' Some believe that Benjamin Franklin composed that famous saying for his best-selling publication *Poor Richard's Almanack*, while others think he's more likely to have recycled it from the wastepaper bin of history. Either way, it neatly reflected the great American's industrious attitude to life.

This same busy frame of mind was also revealed in a satire penned by Franklin in 1784 for the *Journal de Paris*, which is sometimes held up as the first modern call for a form of Daylight Savings Time. Given the vast breadth of his learning, it's possible that Franklin was aware that the water clocks of ancient Rome were adjusted to different timescales from summertime to winter.

After several years living in Paris as the envoy to France of the newly independent American colonies, Franklin finally allowed himself a gentle swipe at the laid-back ways of the locals. He wrote a whimsical essay which appeared in the Economics section of the *Journal*, poking gentle fun at the French fondness for a long lie-in.

Writing anonymously, he calculated that if all the families of Paris who caroused until late at night and then slept until noon would arise with the summer sun six hours earlier, 64 million lb of candle wax would be saved in six months' time. With tongue firmly in cheek, he proposed 'to ring church bells at sunrise, and if that was not enough, let canon [sic] be fired in every street to wake the sluggards'. Adding belt to braces he called for a tax on window shutters that kept out the stirring rays of the rising sun, and for candles to be rationed.

Written with some humorous intent, Franklin's essay was duly dismissed as a joke, and the idea for daylight savings lay in abeyance until the start of the twentieth century when it was revived by a prosperous English builder named William Willett. Legend has it that Willett was returning from an early morning horse ride when he noticed that the blinds were still drawn on many of the local houses while the sun blazed down gloriously.

Willett published a pamphlet *The Waste of Daylight* using his own financial resources. He began canvassing support for an idea he'd had to move the clocks of Britain and Ireland forward in the summer to take advantage of the daylight in the mornings and the lighter evenings. His softly-softly approach was to move the time forward twenty minutes on each of four April Sundays as spring turned to summer, switching them back by the same amount on four Sundays in September. He argued that under his scheme the evenings would remain light for longer increasing daylight recreation time, and this could save huge sums in lighting costs.

Willet's idea found favour with several influential figures including Foreign Secretary Arthur Balfour and the future Prime Ministers David Lloyd George, James Ramsay MacDonald and Winston Churchill. Sadly for Willett, the opposition to his proposal was more powerful still. It included Prime Minister Herbert Asquith, the Astronomer Royal Sir William Christie, the head of the Meteorological Office Sir William Napier Shaw, and the farming lobby. A popular slogan amongst opponents who claimed it would condemn workers to longer hours was that it should be called Daylight Slaving Time. Britain's theatre owners, fearing a drop-off in trade if people were tempted to seek alternative diversions on the brighter evenings, also lobbied against the Bill.

Willett spent a fortune lobbying his idea at Westminster, and his daylight saving plan was tabled for debate by the Liberal MP Robert Pearce who introduced the first Daylight Savings Bill to the House of Commons in February 1908. In support of Willett's scheme, Pearse quoted Ireland's best-known poet of the era, Sir Thomas Moore, who had written: 'The best of all ways / To lengthen our days / Is to steal a few hours from the night.' It was to no avail. The Bill stalled and Willett died in 1915 a disappointed man.

By the time of Willett's death, the Great War had settled into a crippling stalemate for Europe's great powers, becoming a huge drain on lives, manpower and industrial resources. In April 1916 the Germans moved their clocks forward in order to make the most of the daylight and their reserves of coal and oil. The Austrians and the Dutch quickly followed suit, prompting the Westminster Parliament to rush through a bill putting Britain and Ireland on Summer Time.

Following the war's end in 1918, the British Isles stuck with the daylight savings exercise, but many other countries reverted back to standard time until the advent of the Second World War made a necessity out of its reinvention. Today, daylight savings is the rule in countries of higher latitudes. Those savings dwindle to nothing the closer we get to the equator, where day and night are roughly the same length summer and winter.

Two decades after Willett's death, Winston Churchill penned an appreciation for the *Pictorial Weekly* magazine, entitled 'A Silent Toast To William Willett'. In the piece published in 1934, he thanked Willett for the stretch in the daytime that enhanced 'the opportunities for the pursuit of health and happiness among the millions of people who live in this country'.

A decade earlier, the newly independent Irish Free State had to decide whether to continue setting its clocks to British time, or to go it alone. At the time of independence in 1922, Ireland had been on the same time as Britain for only six years. Since time immemorial Ireland had run reasonably well on a series of local times. Hundreds of miles west of London, the noonday sun passed over Dublin's Dunsink Observatory some twenty-five minutes after Greenwich outside London. Meanwhile, town clocks and church bells in the west of Ireland operated with little regard for Dublin time.

This changed with the coming of the telegraph network which required a more synchronised system of timekeeping, and a statute of 1880 made Dublin Mean Time – which was GMT plus twenty-five minutes – the standard legal time for the island of Ireland. This changed again in the wake of the 1916 Rising, when the British authorities decided that Irish Time should be abolished in the name of tighter administration, and Ireland should now run on GMT.

In the spring of 1923 the Oireachtas debated the touchy but pressing matter of whether to continue synchronising the clocks of independent Ireland to British Summer Time, or to go it alone. Senator Butler argued for going it alone, remarking: 'I desire to point out that the Summer Time Act was first forced on this country by the English Parliament, a body legislating for an industrial population. I am quite willing to admit that the Summer Time Act may have conferred some benefits on townspeople [but] the hours of the evening are much more favourable for harvesting operations than those of the morning, which are generally useless for that purpose.'

Butler's view was supported by his fellow senator, Colonel Moore, who said: 'I regret that an Irish government should follow in the footsteps of an English government.' The Colonel insisted that the Oireachtas should throw out the Summer Time Act because it was 'just to please a few lazy fellows in the towns who will not get up early. It is not necessary to put back the clock to make a man get up early; all he has to do is to shake himself, get up an hour earlier or not, or do what he likes. Why should the poor country people be caused all these annoyances to please the fellows in the towns?'

There was further support from Senator J. McLoughlin who argued that the imposition of GMT on Ireland had been a 'violation', and that retaining Summer Time would only compound the injury. Speakers in favour of the Summer Time Bill had claimed that anyone who didn't like it could simply ignore it.

Senator McLoughlin thundered: 'The Government tell us that we can ignore this Bill. It is not very dignified nor self-respecting on the part of the Government to introduce legislation and to invite the country to ignore it. The country cannot ignore it; it is compelled to recognise it. If a farmer wishes to go to a market, a fair, a post office or a town to transact business he must recognise this new time and the result is a welter of confusion in the country, one part going by one time and another by another. If the people want to take advantage of this extra hour of daylight the obvious and natural course for them is to rise an hour earlier. If people are early risers it does not require a Summer Time Bill to get them up. It is only, as Colonel Moore said, the lazy fellows who want to lie in bed until 8 or 9 o'clock in the morning, who want to be coaxed out of bed by legislation and want to be deluded into the belief that 6.30 is actually 8 o'clock. I think it is a ridiculous Bill and a Bill that is simply dignifying humbug and make-belief by legislation, and that the proper name of the Bill should not be the Summer Time Bill, but should be the Lazy Man's Delusion Bill.'

In the Dáil, Walter Cole argued against Summer Time, saying: 'I think it is generally recognised that the workers in towns are more in favour of Summer Time than the people in the country. In the country nearly every one observes the old time. The Summer Time was opposed altogether or ignored. People very seldom recognised it except when they had to catch a train. Nobody was in love with it.'

But it wasn't just the human quality of life for which he feared. In a paroxysm of gombeenism and deeply dodgy maths he asserted: 'The Deputies from the country will bear me out that no self-respecting cow can expect to be milked at 2.30 in the morning, and that according to this new Summer Time in the West of Ireland cattle that in the ordinary course would be milked at 4.30 would now be milked at 2.30. The milk supply of the country is a very important matter. I represent a constituency that is very much interested in the milk supply, and in the Summer Time the people will have to get up practically in the middle of the night in order to milk their cows in time. All these points are deserving of serious consideration.'

Which, of course, they weren't.

One senator to argue in favour of keeping Summer Time was James Douglas, who said that if the Free State was to exist in a separate time zone to Northern Ireland it would harden 'a partition which most of us do not approve or recognise'.

Adding his voice in support of the Bill, Senator Ernest Blythe insisted that the mental exercise of juggling Standard Time, Summer Time and God's Time (ie nature's) could only brighten the intellects of the nation. He argued that the Irish people had adjusted to the 25-minute switch from Irish Time to GMT, even if for a while you were 'continually meeting people in the country who, if it was 9 o'clock were scratching their heads and wondering whether it was 8 o'clock or 10 o'clock. They had great difficulties in making that calculation because of the mental strain it imposed, and they were against the Summer Time for this reason apart from the political objection. I do not think it is a sort of mental strain we should shrink from putting on the people. If they had to think along mathematical lines it would perhaps be good for them.'

Summer Time was adopted as a temporary measure for 1923, and again for 1924, before it was finally made permanent in 1925. In 1968 Britain began observing Standard Time (GMT + 1) on a year-round basis, with no winter change. It was thought the change might help match Britain's timekeeping with that of continental Europe in advance of joining the Common Market (now the EU) in 1973. Ireland followed suit with the change, and then again when the experiment was abandoned after three years.

IRISH WEATHER PROVERBS – LOOKING GOOD

It is a sign of a good day to come when dew is still on the grass two hours after the break of day.

When the stars are far from the moon the day will be beautiful.

When the ash is in leaf before the oak, summer will come early.

A dull cloud and mist mean fine days ahead.

The red moon rises softly over the hill and brings good weather.

A fine spell is due when the sound of the waterfall is hard to hear. A bad spell is due when the sound of the waterfall rings clear.

When the sun's legs are up in the morning and down in the evening, the finest weather is coming.

A white mist halfway up a hill is a sign of muggy weather to come.

A rainbow in the evening means a fine day tomorrow.

When the mist on the new moon dies of thirst, dry weather is in store.

THE FARMER
AND THE FISHER
SHOULD BE
FRIENDS

The Long Campaign for a
Morning Weather Forecast

URING HIS FORTY-FOUR YEARS as a member of the Irish Parliament, Oliver J. Flanagan raged tirelessly against a blacklist of those he insisted were loosening the moral fabric of Irish society, amongst them *á la carte* Catholics, Jews, intellectuals and debauched young women who drank in lounge bars. He is perhaps best known for his memorable statement that 'there was no sex in Ireland before television', but it was in recognition of his overall contribution to the gaiety of the nation that the normally timid 1971 version of the *Sunday Independent* bared its teeth and dubbed him 'the Prime Idiot' of Irish politics.

One of his many colourful contributions to the Dáil record came in a 1953 debate on the state of Irish broadcasting which, in the days before television, meant the meagre fare being provided on the limited radio schedules. Following a discussion (not the first) on whether there were too many English accents on Radio Éireann, Oliver J. raised the issue of the weather forecast which followed Din Joe's Irish dancing show on the evening schedule.

He told the house: 'I have dealt with Din Joe. At the conclusion of the programme one hears the announcement: "Now you will have the weather forecast for farmers and fishermen." How did they link farmers and fishermen? Many fishermen resent being associated with farmers. Many farmers resent being associated with fishermen. I should be glad if the Minister could indicate to the House the reason for this association. I cannot understand how bad fishing weather could affect some of the farmers in my constituency.'

It should be pointed out at this juncture that for all of his time in the Dáil Oliver J. represented the voters in the landlocked midlands constituency of Laois/Offaly.

A government TD pointed out to Deputy Flanagan that the interests of farmers and fishermen were tied together in the Department of Agriculture & Fisheries.

He replied: 'Yes, but I cannot see where the relationship comes in, in regard to weather forecasts. It is definitely resented. The Minister or those in charge may not know that. The smallest farmers are the proudest of all, and they dislike being associated with fishermen. Fishermen dislike being associated with farmers.'

A future Minister for Agriculture & Fisheries, Neil Blaney, reminded him: 'The weather effects both farmers and fishermen.' But Flanagan persisted: 'It is wrong that the weather forecast should be announced in this way. Many small farmers and fishermen resent it.'

'They can go on resenting it, if it is true that they do resent it,' declared the Minister for Transport & Power, Erskine Childers, closing off that avenue of debate.

The world's first radio weather bulletin for consumption by the general public was broadcast on 14 November 1922 by the BBC, with an announcer reading out a script prepared by Britain's Met Office. Remarkably, there was a delay of several months before the BBC began running daily radio forecasts in March 1923.

On New Year's Day 1926, 2RN (pronounced To Éireann) began broadcasting as the official station of the Free State. The opening night was taken up with speeches and an array of live performances intended to showcase the best in Irish traditional and classical music. The schedule for the second night was more typical of the fare to come on a station that broadcast for only three hours each evening, and there, taking its place amongst the singers and the announcers, was the weather forecast.

Before the year was out a government backbencher, George Wolfe, was the first to suggest in the Dáil that the radio weather forecast was not serving the people of Ireland as well as it should. He wanted his own administration to address an issue that was 'very important to country people'. He explained: 'At present the time signal and the weather forecast is given very late, at about ten-thirty. In the country districts it is the custom for the labouring classes to go to bed at about eight-thirty or nine, because they have to get up early. It would be a great advantage to them to have the weather forecast, the time signal, and all that kind of thing, given before the ordinary programme commences, instead of at ten-thirty or later.'

In short, he wanted the weather forecast for the next day to be given at 7.30 p.m. when the station opened, and when the audience was at its greatest. The Minister wouldn't commit either way, preferring to assert that: 'Generally speaking, our programme is a good programme.'

Ten years later, in 1936, the Irish Meteorological Service (Met Éireann) was established to take the Irish weather out of British hands and provide a service better customised to local needs. Except this was very slow in coming. In May 1945 the government pleaded that the war had 'interfered with' the government's plans to improve the service.

Two years later, in 1947, Joseph Blowick of the small farmers' party Clann na Talmhan, revived the calls of twenty years earlier for a radio weather forecast to be broadcast at a time that suited the listener rather than the broadcaster. He urged the Minister that: 'During the spring, summer and autumn months there should be an early morning broadcast, say between nine and ten o'clock, giving the weather forecast. That would be an immense benefit during the haymaking and harvesting seasons to those engaged in agriculture.'

His call was shot down by his own party colleague Bernard Commons who argued: 'I think that we should have, if possible, a news broadcast at nine o'clock every morning. Some Deputy suggested that a weather forecast should be broadcast early every morning in order that everybody might know the prospects of the weather for the day. I think that the man fit to give that forecast would be a remarkable man. Nobody at the present time can forecast the ups and downs of the weather for half an hour, but I think that the average man in the country would appreciate some news, even a few items, at 9 a.m.'

Fianna Fáil's Minister for Posts & Telegraphs Patrick Little informed Commons that the problem wasn't so much the impossibility of giving an accurate forecast for a whole day, but paying for someone to provide it and broadcast it.

And there was the rub. The gathering of both news and weather information was more labour intensive in those days than these, which made it more expensive in real terms. Minister Little

A TO Z
OF IRISH WEATHER

Waterspouts A waterspout is a column of air whirling around a vortex above a body of water. Occasionally seen in Ireland, these mini-tornadoes which connect the water surface to a low cumuliform cloud appear to suck up water, but they don't. The liquid seen in the swirling column consists of water droplets formed by condensation.

Weather Eye To keep a weather eye on someone or something is to be watchful and alert for signs of change or trouble, so that help can be given or evasive action taken if needed.

White Livestock In her 1969 book *The Irish Donkey*, Avril Swinfen argues that white donkeys are rare in Ireland because donkey owners have long believed they are less able to withstand the Irish weather than their darker siblings. She writes: 'White has never been a very popular colour in Ireland with cattle because they were less able to stand up to the rigours of the Irish climate, more susceptible

explained that while he was all in favour of having a better weather service on the wireless, he had to watch the bottom line and ensure that shoddy weather reports would never 'mislead' the Irish public.

The following year Little's Fianna Fáil administration lost office, but the new man in charge of broadcasting quickly came to the same conclusion that the country simply couldn't afford a morning weather forecast.

Minister James Everett told the Dáil: 'We have looked into the matter of a morning forecast and have discussed it with representatives of the Department of Agriculture, but taking everything into to vermin and subject to a complaint called white heifer disease which caused them either to abort or fail to conceive. Since the ass was owned chiefly by country people, the prejudice against the white cow can be assumed to have extended to the white ass, so that the white asses, like white cattle, were most likely gelded if males and seldom bred from if females.' Perhaps lending credence to the notion that white livestock fare poorly in the sun-starved Irish climate, a 1977 study of African cattle by Finch and Western found that dark cattle are far more efficient at absorbing heat from the sun than white cattle, which expend more energy (and therefore burn off more weight) generating body heat.

Wind Chill The wind-chill factor is the perceived decrease in the air temperature felt on exposed skin when a cool wind is blowing. The temperature gauge may give one reading, but for the individual exposed to the breeze, the air can seem much colder than the figures say. Our body heat warms the air immediately touching the skin, forming a shallow but significant layer of insulation around us. A cool wind blows away this layer of self-generated heat, replacing it with colder air, hence the chill.

Winter Rules During adverse weather conditions, especially during the winter, golfers can improve their score, and avoid damaging the course, by applying winter rules or preferred lies. To 'prefer' the lie of the ball, for instance, a player may pick it up and clean it if the local winter rules allow this. Winter rules are not codified under any of the thirty-four Rules of Golf, and each club is allowed to apply its own local ordinances.

consideration, we felt that this is a development that is hardly called for. The station would, of course, have to be opened specially for the purpose, and this would involve considerable expense.'

It would be some years more before the farmers got their early morning weather forecast.

WHEN WEATHER
CHANGED
HISTORY
PART 8

The Mayo Postmaster and the
D-Day Landings, 1944

SOME SAY THAT THE MOST IMPORTANT weather forecast in world history was transmitted in June 1944 from the granite lighthouse-cum-coastguard station that stands in Blacksod Bay on the rugged coast of County Mayo. The report, filed by Blacksod's postmaster Ted Sweeney, delayed the Allied invasion of Normandy by twenty-four hours. Had the landings gone ahead as planned on 5 June rather than 6 June, the outcome may have been very different.

By the end of 1943 the huge build-up of Allied troops in the south of Britain alerted the Germans occupying France that an invasion was coming. The Nazi commander in charge of the coastal defence requested and received reinforcements. The most senior new addition was Field Marshal Erwin Rommel, who spotted that the Germans' Atlantic Wall fortifications were strong around the key ports, but puny along the beaches.

Under Rommel's direction the defences were strengthened along the vulnerable stretches of beach, but the Germans left an Achilles' heel exposed that would cost them dearly. The Nazis assumed that the Allies would attempt to land their troops at high tide to place them as far up the beach as possible. With this in mind, they built their heaviest fortifications on the foreshore area, the part of the shore that lies just above the high-water mark. Second-guessing the Germans, the Allies decided to land at low tide in the hope the defenders would have to shift their firing positions.

While US General Dwight D. Eisenhower was poring over the final details of the D-Day landings, postmaster Ted Sweeney and his wife were going about their daily chores in far distant Mayo, blissfully unaware that the fate of the western world was about to be decided. One of the couple's duties was to file an hourly weather report to London. During the early years of the Irish Free State, the British Meteorological Office in London continued to provide expertise to, and process data from, the network of weather stations around Ireland. The Irish Meteorological Service (now Met Éireann) was established in 1936 to take over administration of information gathering from the British, but it remained in the mutual interest of both countries to share their weather data.

For Eisenhower, who was Supreme Commander of the invasion forces, the opportunity for a mass landing was limited to only a few days in each month as a full moon was required to ensure both a spring tide and light for the pilots. Eisenhower had tentatively selected 5 June as the date for the assault. However, when 4 June arrived conditions were clearly unsuitable for a landing. High winds and lurching seas made it impossible to launch landing craft, and low clouds would hinder aircraft from finding their targets. With the weather foul and the forecast predicting no let-up, the Germans switched off. Troops were stood down, while some of their officers took time off. Rommel, for instance, took leave to attend his wife's birthday. Eisenhower, on the other hand, believed that the conditions on 5 June, while far from ideal, would enable him to strike.

Speaking to *The Irish Times* in June 1994, fifty years after the event, Ted Sweeney recalled: 'I was sending an hourly report for the 24 hours day and night. It had to be phoned in to London. We got a query back. They asked for a check. "Please check and repeat the whole report." I went to the office and checked the report and repeated it again. I just wondered what was wrong. I thought I had made some error or something like that. They sent a second message to me about an hour after to please check and repeat again. I thought this was a bit strange so I checked and repeated again.'

On 4 June Ted's report to London indicated that a cold front was moving with considerable speed across Ireland. His information, which he was asked to reconfirm several times, suggested that heavy rain and force seven gales would hit the English Channel the following day, which would have played havoc with the invasion force. On a brighter note, the same report indicated that a short window of relative calm weather would follow in the wake of the storms. Eisenhower delayed D-Day by twenty-four hours and turned the tide of the war.

The tale of how Ted Sweeney, the postmaster of Blacksod Bay, played a key role in the history of the world, has grown legs. One legend has it, for instance, that a high-ranking officer of Eisenhower's Supreme Headquarters Allied Expeditionary Force (SHAEF) phoned him directly the day after the successful landings to congratulate him on his observations. Some experts insist that the reports filed to London from the weather station on Valentia Island in County Kerry were no less important than the ones from Blacksod, as each confirmed the other. The German forecasters, who were unable to access the data from the west of Ireland, fatally misplaced the approaching cold front of 4 June some 240 miles west of where it actually was.

The Free State's decision to remain neutral during the Second World War was resented by Britain's leadership and most of its people. This may account for the fact that many accounts of Operation Neptune (the landing phase of Operation Overlord) ignore completely the role played by the weather stations in Mayo and Kerry. These versions generally, but not very convincingly, credit the Royal Navy frigate HMS *Grindall* sailing in the mid-Atlantic as the sole source of the vital information.

Speaking half a century after the event, Ted Sweeney made no claim that he'd been personally congratulated by Eisenhower's headquarters, but he was satisfied that the role he'd played in deciding the war was valid and valuable. He recalled: 'Months afterwards, perhaps even a few years, I was told by one of the Met Office crowd that the reason they took heed of that particular report for Blacksod Point was that it was the first indication they got of a major change in the weather. It never dawned on me that this was the weather for invading or anything like that. When I checked the report, I said "Thanks be to God I was not at fault anyway". I had done my job and sent over a correct reading to London.'

Had Eisenhower postponed the invasion beyond 6 June, he would have had to wait another fortnight for the next viable window of opportunity. As it turned out, that window, from 19 to 22 June 1944, unleashed what Britain's Prime Minister Winston Churchill described as 'the worst channel storm in forty years'.

By then, happily, the clock was running down fast on Hitler's Thousand-Year Reich.

The Big Freeze of 1947 in Numbers

19

On 19 January 1947 Ireland was blitzed by an Arctic snowstorm of brutal intensity. The country was covered with a thick blanket of snow the likes of which few had ever seen.

5

Over the next two months the number of major blizzards to hit the country shot up to five, the worst of which occurred on 24 February.

-14

The lowest temperature recorded was -14 °C, but it rarely rose above freezing point for weeks on end. The result was that each fresh snowfall formed another imposing layer on the ones previously laid down.

1847

With virtually every road impassable, and the transport system brought to a standstill, the flow of food supplies was reduced to a dribble. As they shivered under siege in their homes, the Irish people had long hours to dwell on the fact that it was the hundredth anniversary of Black '47, the worst year of the Great Famine. With crops and livestock dead in the fields, an ancient fear of famine reasserted itself.

16

Both humans and animals were forced to subsist on meagre rations. With livestock fodder scarce, the price of hay soared from a normal rate of some £4 to £5 per ton, to £16, where it could be got.

34,000

Stricken with its own fuel supply problems, Britain suspended exports of coal, at a time when Ireland's turf stocks lay deep under snow, or too sodden for use. Britain relented and released small quantities of coal and coke as relief aid, but these were dwarfed by a United States' pledge of 34,000 tons of emergency coal.

14

After two months of deep freeze, Saint Patrick's Day heralded the thaw the nation prayed for. However, a sudden jump in temperature produced a massive deluge of meltwater, reportedly flooding some urban areas to a depth of 14 feet.

NO RESPECTER
OF FAME

Ryan's Daughter, Led Zeppelin
and the Irish Weather

IN HIS SEMINAL WORK *Modern Painters*, the celebrated Victorian art critic John Ruskin coined the term 'pathetic fallacy' for instances where inanimate objects are treated as if they had feelings, motivations or intentions. The word 'pathetic' relates to 'pathos' or 'empathy' and doesn't carry a negative slant.

To illustrate what he meant by pathetic fallacy, Ruskin quoted a couplet from Charles Kingsley's poem *The Sands of Dee*: 'They rowed in across the rolling foam / The cruel, crawling foam.'

Ruskin pointed out: 'The foam is not cruel, neither does it crawl. The state of mind which attributes to it these characters of a living creature is one in which the reason is unhinged by grief.'

As far as the master critic was concerned, this was all fine if it was done well. He explained: 'So long as we see that the feeling is true, we pardon, or are even pleased by, the confessed fallacy of sight which it induces: we are pleased, for instance, with those lines of Kingsley's, above quoted, not because they fallaciously describe foam, but because they faithfully describe sorrow.'

Ruskin was putting a name on something that artists and writers and ordinary folk had been doing since the dawn of time, and nothing across the ages has had more pathetic fallacy projected on it than the weather.

Frank McCourt casts the Irish weather as a lead villain in his 1996 Pulitzer-winning memoir *Angela's Ashes*. As a frail schoolchild, the young McCourt wins the approval of his teacher by penning an essay entitled 'Jesus and the Weather'. In the essay he argues that if the Baby Jesus had been born in Limerick rather than balmy Palestine he'd never have survived long enough to become a child prodigy. The pathetic fallacy of *Angela's Ashes* presents the Limerick weather as cruel and tyrannical. At one point McCourt says: 'Out in the Atlantic Ocean great sheets of rain gathered to drift slowly up the River Shannon and settle forever in Limerick.' Elsewhere, the dank oppressive atmosphere would seep into the mind and the lungs and ultimately into the soul: 'It created a cacophony of hacking coughs, bronchial rattles, asthmatic wheezes.'

'Above all,' he concludes, 'we were wet.'

If the weather could be used to show Limerick as the Sick Man of 1930s Ireland, it was also famously used to portray Ireland as the Sick Man of Europe in David Lean's epic *Ryan's Daughter* starring Robert Mitchum, Sarah Miles and John Mills.

David Lean came to Ireland fully intending to use the County Kerry weather as an uncredited lead character in his movie, which tells of an affair between a married local woman and a British officer in a dysfunctional rural community. Lean had already subdued the jungles of Sri Lanka to film *The Bridge on the River Kwai,* and the deserts of Jordan and Morocco for *Lawrence of Arabia.*

Coming from the neighbouring island of Britain, Lean must have felt that he could comfortably coach and cajole the starring performance he wanted from the Kerry weather. The Kerry weather had other ideas, and there was only ever going to be one winner.

Lean arrived outside the town of Dingle with everything planned in his mind. It was 1969 but the Dingle Peninsula was barely changed from the way it looked in 1916, the fictional year of the movie. The director's plan was to have everything in the can inside three months, and get on with his next project. Instead, fourteen months later the production crew finally pulled out of Dingle, frayed and ratty from their extra year of confinement.

The Irish weather had wreaked havoc with Lean's best-laid plans. His vision called for many of the love scenes between Rosy (Sarah Miles) and Doryan (Christopher Jones) to be played out on a beach bathed in golden sunshine. But as anyone who's ever spent a summer holiday in the south of Ireland could have told him, the Kerry weather doesn't 'do' sunshine, unless the mood takes it.

After endless hours, days and weeks going stir-crazy in caravans, gazing out at the pelting rain, the crews were overjoyed to learn that a stretch of beach had been found in sunny South Africa that could pass for Dingle Bay. They were off like a light.

After an expensive round-trip to Apartheid Africa, the crews arrived back to shoot one of the film's pivotal episodes. The script called for a tumultuous storm blowing in from the sea as a suitable backdrop for a scene where crates of rebel arms are washed up on shore. Asked to act merely natural on cue, the Irish weather got all shy and docile. The demented camera crews had yet more waiting to do, and in the end Lean was forced to splice together bits from five different storms.

Journalist Helen McCormack of the London *Independent* told later of how, echoing the famous words of Winston Churchill, the capricious weather had taken upon itself to fight the attempted love-making of the two leads not just on the beaches but in the fields too. She wrote: 'Shooting was planned in a field of bluebells after the lovers dismount from horses and begin a passionate embrace. Filming was hampered by … a surprise downpour … addressed by hiring the village hall, where the crew literally grew a field, with grass seeds planted under specially created humid conditions and butterflies and birds bought into the hall.'

David Lean left for Hollywood more than a year over schedule and £4 million over a £9 million budget. That was the price he paid for giving the Irish weather less than its due respect. And after all that, when it finally opened in late 1970 *Ryan's Daughter* was mauled by the critics as the worst thing he'd ever done.

Three years later the Irish weather handed out another battering to the world of international showbiz, but this time with somewhat happier results. In 1973 Led Zeppelin were the biggest rock act on the planet. The outfit had found massive success with their first four albums entitled, imaginatively, *Led Zeppelin I, II, III* and *IV*. They wanted their fifth collection to be a departure, and to this end gave it a proper name – *Houses of the Holy*. They decided that the sleeve should also give notice that they were moving on, towards a more hi-tech, space-age sound. They summoned the hippest design team of the day, Aubrey Powell and Storm Thorgerson of Hipgnosis. The pair were ushered in to present their ideas before a judging panel of singer Robert Plant, guitarist Jimmy Page and the band's scary Svengali Peter Grant.

It being the 1970s, the band hadn't bothered to give the Hipgnosis team a hint of their new musical direction to go on, or even a tentative album title. Left completely in the dark, the designers presented the stellar jury with just two concepts. One involved a photo shoot in Peru, and the other involved lots of children sprawled on Antrim's Giant's Causeway in homage to Arthur C. Clarke's sci-fi classic *Childhood's End*. Told that either option would be obscenely expensive, Manager Grant snarled back: 'Money? We don't f**king care about money! Just f**king do it!'

And that's how one of the most iconic album covers in the history of rock began life, with the London-based design team opting for the short hop over the long haul.

Powell set out for Northern Ireland with two children and a crew of photographers, lighting and make-up people. And, to their seeming surprise, it poured and poured and poured.

Powell had planned to shoot the cover shot on colour film, and for ten days on the trot the crew assembled on the Causeway morning and night to capture the desired eerie light of dawn and dusk. But the Antrim rainclouds had settled in for the duration. As Powell later recalled: 'It promptly rained for ten days straight. [Eventually] I shot the whole thing in black-and-white on a totally miserable morning pouring with rain.' Although the cover looks like a single wide-frame shot with lots of children, it's actually a collage of thirty frames with the same two children multitasking.

Powell continued: 'Originally I'd intended the children to be gold and silver, but because I shot in black-and-white on a grey day, the children turned out very white.' Failing to compensate for the unwanted new skin colour, the artist tasked with hand-tinting the children's bodies gave them a day-glo purple tinge.

It was the happiest of accidents. As Powell later recalled: 'When I first saw it I said "Oh My God!" Then we looked at it and I said, "hang on a minute – this has an otherworldly quality". So we left it.'

The following year the sleeve of *Houses of the Holy* was nominated for a Grammy in the Best Package category, while in 2003 it was rated Number Six on VHI's 50 Greatest Album Covers of All Time list.

NIGHT APPEARS TO FALL BY MIDDAY

Smog Menaces The Capital

IN THE LATE 1980s Met Éireann added an unwelcome new strand to its evening weather forecasts – the smog alert. Most of the warnings applied to Dublin and, indeed, to particular built-up areas of what could be called the old city, with Ballyfermot and its tight-packed terraces of labourers' cottages suffering particularly badly. The smog was at its worst on short, cold winter days when thousands upon thousands of chimneys pumped out shrouds of coal smoke that suffocated communities like a filthy dank blanket.

The notoriety of Dublin's smog spread far and wide. In January 1989 *The New York Times* ran a major feature melodramatically headlined 'Fair Is City But Foul Is Air When Smog Creeps In'.

The piece opened like a scene from a Penny Dreadful thriller: 'The smog creeps menacingly through doors and windows here. It attacks throats and lungs. It sometimes invades Dublin to such a degree that night appears to fall by midday.'

The writer went on to report that surgical facemasks were being distributed to counter the smog that was 'seven times above pollution levels considered worrisome by the World Health Organization'.

The New York Times explained that the poisonous smog was caused by a temperature inversion. It said: 'Unlike the photochemical smogs that shroud some of the world's large metropolises, the haze here is smoke pollution. More than 80% of the problem stems from the 350,000 tons of coal Dubliners burn each year in their cozy fireplaces. Most of the coal is bituminous, which produces much smoke and ashes. The Government says that for every ton burned in an open grate, more than 100 pounds of smoke, along with sulfur dioxide, is produced. On cold, windless nights when this city experiences a temperature inversion, in which warm air traps cooler air near the earth's surface, preventing the normal rising of surface air, the cool air hovering over the ground can trap the smoke spiraling from chimneys. The result is smog.'

As far as we know, the word 'smog' first appeared in print in the *Los Angeles Times* in January 1893, in an anonymous feature attributed to 'a witty English writer'. It appears to have entered common currency after it was used in a 1905 public health speech by Dr Henry Antoine Des Voeux, who was quoted in London's *Daily Graphic* newspaper as remarking that: 'It required no science to see that there was something produced in great cities which was not found in the country, and that was smoky fog, or what was known as smog.'

In January 1989 *The New York Times* reported that the Irish government was attempting to tackle the smog menace by imposing

Ten Weather Forecasting Creatures from Michael Gallagher

Donegal postman Michael Gallagher is Ireland's best-known amateur weather forecaster. As man and boy he has absorbed the weather wisdom of the natives of the Bluestack Mountains in the northwest corner of the island. He says: 'All along the western seaboard the people never forgot how to read the signs. When I was a child, my mother would always wonder why I took so long going to the shops for messages, but I was stopping to observe nature and to ask older people how they knew what the weather was going to do. What I learned early on is that animals are more intelligent than people when it comes to reading the weather signs. We can learn a lot from observing them.'

Here are ten of the many creatures that Michael looks to for guidance on the weather to come.

Badger When autumn comes, if the badger stockpiles leaves and branches at the mouth of his sett, this can be taken as a sign that winter will arrive early.

Frog If the coming summer is going to be a washout, frogs will lay their frogspawn on dry-ish level ground in anticipation that the spawn will be well watered. If the summer is going to be dry and sunny, they will place their spawn close to a stream or a pond to ensure there is adequate water for their offspring.

Hen Sensing that the weather is going to stay fair and settled for a spell, hens will stray further than normal from their roosts, foraging in open meadows.

Horse The horse shares a bad-weather homing instinct with the hens. If it's going to stay fine the horse will venture to higher ground to graze, but if the weather is about to take a turn for the worse it will stick close to home, perhaps giving the occasional anxious glance to the horizon.

Sheep Although sheep have a reputation as being amongst the dumbest of God's creatures, they are smarter than the average human when it comes to detecting subtle changes in the atmosphere. When they frolic in the autumn, winter or early spring, it is a sign of snow on the way. When they bunch tightly together, it foretells high winds coming. When they head for the hills in late spring without being driven there, it is a sign that the worst of the weather is over and summer is on the way.

Dolphin Fishermen don't like to see dolphins appearing near Irish coasts early in the year, as this betokens bad conditions to come. Dolphins arriving in the early summer is a good sign.

Midge When these biting pests muster in dense clouds, rain is on the way.

Flying Ant Like the midges, when these appear in great numbers, rain is due.

Spider If spiders weave their webs in the shelter of doorways and windowsills, expect bad weather. If they spin their webs on the tops of rushes and bushes in the early spring, good weather is due.

Cat Felines exhibit various forms of behaviour that can indicate the weather to come. Fishermen take particular notice when a cat starts clawing at the legs of a table or chair, as this is often a warning that high winds are on the way.

Michael Gallagher is the author of the books *Traditional Weather Signs* and *Remedies And Cures of Bygone Era*.

'smoke control' zones on the worst-hit areas, where householders could use only authorised 'smokeless' fuels. The government campaign was backed with grants to convert from smoky coals, and encouraged with posters, and the radio and TV slogan: 'Stop smog – switch to smokeless.' Meanwhile, Dublin's chemist shops did a roaring trade selling disposable 'smog masks' at the substantial price of 95p a go.

As *The Irish Press* pointed out as the smog started to become an annual health hazard, the grants to stop using coal contradicted a series of earlier grants encouraging the public to switch to coal a few years earlier. In 1985 the *Press* observed: 'The oil price explosion of the 1970s led to many householders switching back to coal fires. Indeed, many found their central heating bills so excessive that they insisted on local authorities providing them with chimneys so they could use the cheaper fuel.'

Environment Minister Mary Harney's campaign to end Dublin's annual smog season eventually came good, but at the time of *The New York Times'* visit, Alan Shatter of the Fine Gael opposition derided her efforts.

As *The Irish Press* noted: 'Mary Harney was accused yesterday of spending as much on a poster advertising campaign about smoke control as on grants to help people switch to smokeless fuels. FG's Alan Shatter said it was outrageous to spend £120,000 on the poster campaign – which he claimed could not be seen by the public due to the smog.'

THE MINISTER
FOR SNOW

Sound Reasons For Holding
Summer Elections

THE RTÉ TELEVISION NEWS IN 2009 carried a report on a flooding disaster that had left large tracts of the country under water. Having surveyed some of the damage from a helicopter, Environment Minister John Gormley lamented: 'You can't legislate for acts of God.'

Two years later, shortly after losing his seat in the general election of February 2011, a reflective Gormley told RTÉ TV's *Weather Permitting* documentary: 'A lot of people said that from the time we [the Green Party] entered government to the time we left, it didn't stop raining. And the weather was extremely bad. Obviously that was entirely coincidental ... This coincided with the financial crisis. We had very severe weather, we had floods, we had very severe snow and ice subsequently. And then we had people going through the trauma of a financial meltdown. All of that combined, obviously, led to a feeling of discontent, a feeling of dissatisfaction. It's up to the electorate to afterwards cast their judgment on that – and as you know they did.'

The former Minister clearly was not trying to lay his party's electoral wipeout at the hands of a bad run of weather, but he was acknowledging that even the ship of state can be buffeted by the twists and turns of the elements.

As John Gormley noted, politicians can't legislate for acts of God, but they generally try their best to engineer the dates of elections in the hope of catching the weather at its best-behaved. The terms 'fairweather voters' and 'sunshine vote' have their basis in generations of political folk wisdom that the plain people are not only more likely to come out and vote in good weather, but they may be better disposed to their ruling class as the 'feelgood factor' comes into play.

Surprisingly little scientific research into this has been carried out in these islands, but one study, by academics at Plymouth University, has argued that a British general election held in the first week of June would of itself increase turnout by 3 per cent over one held in the first week of April.

Under the Local Government Act of 2001 Irish local elections must be held every five years in the month of May or June. The official reason for specifying these bright-lit weeks is to tie in the local elections with European Parliament voting across the Continent, but the reason the Continent votes at this time is to try for the best weather.

From long before there was a European Parliament, Irish voters have been called to the polling stations more often than not in May and June. Since the foundation of the State, general elections have fallen overwhelmingly in the months of May and June. Just one has been called for January (1933), while none has ever taken place in dark December.

Ireland's politicians know, almost instinctively, that if they call a general election in the depths of winter, they risk getting a jolt of any discontent stalking the land. Sometimes, however, they get a sharp reminder of what the British Prime Minister Harold Macmillan meant when he was asked what is most likely to blow governments off course. He answered: 'Events, dear boy, events.'

In January 1982 the recently installed Fine Gael/Labour government led by Garret FitzGerald unveiled a savage debut budget. In doing so, FitzGerald failed to heed the warnings of independent socialist Jim Kemmy, who'd threatened that he would resist the removal of subsidies on food and the imposition of VAT on clothing. FitzGerald patiently explained that if you exempted children's shoes from VAT, selfish small-footed women might be tempted to diddle the State of its rightful revenue. True to his word, Kemmy pulled the plug and the coalition collapsed in a state of numbed shock after just seven months in office.

In the ordinary course of events, no government would ever have called a general election at that point in time. The country was just emerging from three weeks of blizzards which had brought transport to a standstill, killed livestock on a large scale, caused massive food and fuel shortages and power cuts, and generally reduced the land to a state of chaos. With the Taoiseach out of the country for much of the crisis, the Labour leader and Tánaiste Michael O'Leary had taken charge of the crisis management. For his troubles, he was nicknamed 'The Minister For Snow' (and not in a good way).

The electorate emerged to vote on 18 February 1982, grumpy and resentful like a hibernating bear roused from its slumber with a sore head. The coalition was dumped out after just over half a year in office. As the political reporter Charlie Bird reflected thirty years later: 'That weather crisis severely damaged Michael O'Leary.'

Nine years earlier, Taoiseach Jack Lynch had given an object lesson as to why a sitting government should never call a general election in the depths of winter unless it's absolutely necessary.

In January 1973 Lynch's Fianna Fáil administration still had more than a year to run on its five-year term of office, but the Taoiseach believed he could catch the opposition, and indeed the electorate, napping with a snap election ahead of a closing deadline.

There were mere weeks to go before, for the first time, the voting age in the State was to be lowered from twenty-one to eighteen, which meant that over 140,000 new voters were just about to be enfranchised. However, by calling a quickie campaign of just three weeks' duration, Lynch was able to beat the introduction of the new electoral register by a fortnight.

Fianna Fáil feared the verdict a fresh generation of young voters might pass on the party, which by then had been in power with just

A TO Z
OF IRISH WEATHER

Xbox. Your Xbox can now supply you with weather information for Ireland courtesy of the Xbox Media Centre. The service is entitled 'instant global weather on demand'. Or you could just Google it.

Year Without A Summer The year 1816 was variously known across the Northern Hemisphere as the Year Without A Summer, Poverty Year, Eighteen Hundred And Froze To Death and other names indicating a year of abnormally low temperatures and heavy rains. The failure of the wheat, oats and potato crops brought famine to the north and southwest of Ireland, while failed harvests across Europe brought widespread starvation and rioting. The devastating weather has been attributed to a massive eruption of Mount Tambora in the Dutch East Indies, now Indonesia, in April 1815. The eruption, which cloaked the Earth from the sun with a veil of soot, has been given a ranking of seven on the Volcanic Explosivity Index, earning it classification as a super-colossal event. The freakish summer of 1816 also gave birth to two short breaks for thirty-five years on the trot, prompting one commentator to note: 'They look less like the Soldiers of Destiny and more Dad's Army.' The Minister for Posts & Telegraphs Gerry Collins might have chosen his words better when he said of his leader, Jack Lynch: 'We reject the implication that a Taoiseach who has been in office a mere six years is a tired old man.'

Student leader Pat Rabbitte said he was sickened by the move, remarking: 'The democratic right of almost 200,000 young people is to be sacrificed in the party political interests of Fianna Fáil.'

In the end, it wasn't the young people of Ireland who saw off the government, but the weather may well have tipped the result. Snow and ice covered the campaign trail across much of the land and the voters gave Lynch and his strategists the cold shoulder.

---

one of the greatest fictional freaks of world literature. The teenage Mary Wollstonecraft Godwin arrived at Lake Geneva in May that year with her husband-to-be, the poet Percy Shelley, to stay with a group of bohemians that included the glamorously rakish poet Lord Byron. Years later, she recalled: 'It proved a wet, ungenial summer and incessant rain often confined us to days in the house. "We will each write a ghost story," said Lord Byron.' Mary didn't write a ghost story, but she did write the groundbreaking gothic novel *Frankenstein*.

Zoo. A bonus attraction of Dublin Zoo during the bitterly cold winter of 1963 was ice skating on the frozen lake. A decade earlier in 1954 the maintenance of tropical beasts in the unkind Irish climate was proving a burden in a land wracked by poverty, stagnation and emigration. The zoo invited individuals or firms to adopt 'lonely animals' by sponsoring their feed and upkeep for a year. An elephant could be maintained for around £200, a lion or tiger for about £100, and a puma or leopard for roughly half that again. One holiday company with a beaver on its logo answered the call by saying they'd like to sponsor one of these wet-look rodents. A zoo spokesman replied hopefully: 'At the moment we have no beaver but we hope to go get one.' The same official expressed concern that, with horsemeat so scarce and pricy, the zoo might soon have to start feeding the lions with whale meat.

Selected Bibliography and Sources

Butel, Paul, *The Atlantic* (Routledge)
Chadwick, Nora, *The Celts* (Pelican)
Corless, Damian, *Party Nation* (Merlin)
Davis, Norman, *The Isles* (Macmillan)
Dickson, David, *Arctic Ireland* (White Row Press)
Ferguson, Patrick, *Troubled Waters* (Nonsuch)
Gallagher, Michael, *Traditional Weather Signs* (Michael Gallagher)
Gray, Peter, *The Irish Famine* (Thames & Hudson)
Kearns, Kevin C., *Ireland's Arctic Siege* (Gill & Macmillan)
Knowlson, T. Sharper, *The Origins of Popular Superstitions And Customs* (Senate)
Lewis, Peter & Pearlman, Corinne, *Media & Power* (Camden Press)
McWilliams, Brendan, *Weather Eye* (Lilliput Press)
Swinfen, Averil, *The Irish Donkey* (Lilliput Press)

Irish Independent
The Irish Times
The Freeman's Journal

National Library, Dublin
Met Éireann website: www.met.ie
Oireachtas website: www.oireachtas.ie